Handbook of Post-Processing in Additive Manufacturing

Handbook of Post-Processing in Additive Manufacturing is a key resource on post-processing treatments available for additive manufactured products. It provides broad coverage of the theory behind emerging technology, material development, functional characterization, and technical details required to investigate novel applications and methods and put them to use.

The handbook presents experimental breakthroughs of novel methodologies that treat additively manufactured parts, which are suitable for demanding engineering applications. This handbook emphasizes the various types of post-processing technologies that can effectively eliminate the inferiorities of additively manufactured components. It also provides a collection of key principles, literature, methodologies, experimental results, case studies, and theoretical aspects of the different types of post-processing techniques, along with different classes of materials and end-applications.

This book is an ideal reference for libraries and post-graduate courses as well as the professional market, including, but not limited to manufacturing, mechanical and industrial engineering, and materials science.

Handbook of Post-Processing in Additive Manufacturing

Requirements, Theories, and Methods

Edited by
Gurminder Singh, Ranvijay Kumar,
Kamalpreet Sandhu, Eujin Pei,
and Sunpreet Singh

CRC Press
Taylor & Francis Group
Boca Raton London New York

CRC Press is an imprint of the
Taylor & Francis Group, an informa business

Designed cover image: Shutterstock

First edition published 2024
by CRC Press
6000 Broken Sound Parkway NW, Suite 300, Boca Raton, FL 33487-2742

and by CRC Press
4 Park Square, Milton Park, Abingdon, Oxon, OX14 4RN

CRC Press is an imprint of Taylor & Francis Group, LLC

ISBN: 978-1-032-23172-3 (hbk)
ISBN: 978-1-032-23173-0 (pbk)
ISBN: 978-1-003-27611-1 (ebk)

DOI: 10.1201/9781003276111

Typeset in Times
by Newgen Publishing UK

Contents

Preface

This *Handbook of Post-Processing in Additive Manu-facturing: Requirements, Theories, and Methods* aims to present various experimental findings on novel methodologies to treat additively manufactured parts, suitable for demanding engineering applications. This book emphasizes the different types of post-processing technologies that can effectively eliminate the inferiorities of additively manufactured components. This book provides a collection of critical principles, literature, methodology, experimental results, and theoretical aspects of different post-processing techniques. In particular, the book is designed to cover the research fraternity's significant contribution worldwide for other materials and end applications classes. The text is a crucial source of information on the post-processing treatments of additively manufactured products. This ground-breaking book provides broad coverage of the theory behind this emerging technology, material development, functional characterization, and the technical details required for readers to investigate the novel applications of the involved methods for themselves.

About the Editors

GURMINDER SINGH

Gurminder Singh is an assistant professor at the Mechanical Engineering Department, Indian Institute of Technology Bombay, India. He worked as a postdoc researcher in the Mechanical and Materials Engineering School at University College Dublin (UCD), Dublin, Ireland. Before joining UCD, Dr Singh was a Postdoc fellow at SIMAP Lab, University of Grenoble Alpes, France. His research focused on the experimental and simulation studies of the extrusion 3D printing of metals. Dr Singh was awarded by Gandhian Young Technological Innovation Award for developing a 3D-printing method for fabricating patient-specific stents in 2020. Dr Singh obtained his PhD from the Indian Institute of Technology in Delhi, India. His PhD work focused on developing low-cost rapid manufacturing techniques using ultrasonic vibration-assisted sintering and 3D printing. The *3D Print* website featured his PhD work in the February 2020 edition. He has completed his ME from Thapar University, India. He has worked on comparing modern controllers for the swing-up and stabilization of inverted pendulums and was a team member in the Indo-Swedish project on exoskeletons. He completed his bachelor's from Guru Nanak Dev Engineering College, Ludhiana, India, in Production Engineering. Dr Singh also has research fellow experience at IIT Delhi and production engineer experience at Shakunt Enterprises, India. He has published over 35 articles in peer-reviewed international journals, conference proceedings and book chapters. He has published in leading journals such as *Additive Manufacturing, Materials Science and Engineering: A, Journal of Manufacturing Processes, Rapid Prototyping*, etc.

RANVIJAY KUMAR

Ranvijay Kumar is an Assistant Professor at University Centre for Research and Development, Chandigarh University. He has received PhD in Mechanical Engineering from Punjabi University, Patiala. Additive manufacturing, shape memory polymers, smart materials, friction-based welding techniques, advance materials processing, polymer matrix composite preparations, and reinforced polymer composites for 3D printing, plastic solid waste management, thermosetting recycling and destructive testing of materials are the research interests of Dr Kumar. Dr Kumar won the prestigious CII MILCA award in 2020. He has co-authored more than 43 research papers in science citation indexed journals, 38 book chapters, and has presented 20 research papers in various national/international level conferences. He has contributed extensively in additive manufacturing literature with publications appearing in the *Journal of Manufacturing Processes, Composite Part: B, Rapid Prototyping Journal, Journal of Thermoplastic Composite Materials, Measurement, Proceedings of the Institution of Mechanical Engineers, Part C (iMeche Part C), Proceedings of the Institution of Mechanical Engineers,*

Part H: Journal of Engineering in Medicine, Journal of Thermoplastic Composite Materials, Materials Research Express, Proceedings of the National Academy of Sciences, India Section A: Physical Sciences, Journal of Central South University, Journal of the Brazilian Society of Mechanical Sciences and Engineering, Composite structures, CIRP Journal of Manufacturing Science and Technology etc. He is the editor of the book *Additive Manufacturing for Plastic Recycling: Efforts in Boosting A Circular Economy* published by CRC Press (Taylor and Francis).

KAMALPREET SANDHU

Kamalpreet Sandhu is an Assistant Professor in the Product and Industrial Design Department at Lovely Professional University, Phagwara, Punjab, India. He is also editor of various books: *Sustainability for 3D Printing, Revolutions in Product Design for Healthcare, Food Printing: 3D printing in Food Sector and 3D printing in Podiatric Medicine*. He also acts as an Editorial Review board member for the *International Journal of Technology and Human Interaction (IJTHI), Advances in Science, Technology and Engineering Systems Journal (ASTESJ)* and also as a review editor for the *Frontiers in Manufacturing Technology* section "Additive Processes".

EUJIN PEI

Eujin Pei is a Reader in additive manufacturing at Brunel University London. He is the Associate Dean for the College of Engineering, Design and Physical Sciences and Director for the BSc Product Design Engineering programme. Eujin is a Fellow and Council Member of the Institution of Engineering Designers (FIED), a Chartered Engineer (CEng), a Chartered Environmentalist (CEnv) and a Chartered Technological Product Designer (CTPD). His research focuses on design for additive manufacturing, and he chairs national (BSI AMT/8) and international standardization committees (ISO TC261/WG4). Eujin is also the Editor-in-Chief of the *Progress in Additive Manufacturing Journal*, published by Springer Nature.

SUNPREET SINGH

Sunpreet Singh works in additive manufacturing, composites, biomaterials, process control, and sustainability. He is an Adjunct Professor and Visiting Researcher associated with Chandigarh University, India and the National University of Singapore, Singapore. He has published about 250 research articles in prominent journals. He received his PhD in Mechanical Engineering from Guru Nanak Dev Engineering College, Ludhiana, India. His area of research is additive manufacturing and the application of 3D printing to develop new biomaterials for clinical applications. He has contributed extensively to additive manufacturing literature with publications appearing in the *Journal of Manufacturing Processes, Composite Part: B, Rapid Prototyping Journal, Journal of Mechanical Science and Technology, Measurement, International Journal of Advance Manufacturing Technology,* and the *Journal of Cleaner Production*. He has authored 250 research papers and 27 book chapters. He is also the Guest Editor of three journals: a Special Issue of "Functional Materials and Advanced Manufacturing", *Facta Universitatis*, Series: *Mechanical Engineering* (Scopus Indexed), *Materials Science Forum* (Scopus Indexed), and Special Issue on "Metrology in Materials and Advanced Manufacturing", *Measurement and Control* (SCI indexed).

1 Additive Manufacturing Techniques

Fundamentals, Technological Developments, and Practical Applications

Jasvinder Singh[1], *Ravinder Pal Singh*[2], *and Pulak Mohan Pandey*[3]
[1]Department of Production and Industrial Engineering, Punjab Engineering College (Deemed to be University), Chandigarh, India
[2]Department of Mechanical Engineering, MMEC, Maharishi Markandeshwar (Deemed to be) University, Mullana, Ambala, India
[3]Department of Mechanical Engineering, Indian Institute of Technology Delhi, New Delhi, India

CONTENTS

DOI: 10.1201/9781003276111-1

1.1 INTRODUCTION

Additive manufacturing (AM) is referred to as the process of fabrication in which complex, complicated, and customized geometrical structures are fabricated by using the layered deposition methodology. The initiation of the process occurs with computer-aided-design (CAD) models, which represent actual shape of the physical part needed to be fabricated. The digital sliced information of a CAD model is further transferred to the printing device in which layer-by-layer material development takes place to fabricate the final part. Additive manufacturing is also known as rapid prototyping as well as three-dimensional (3D) printing. The developer of the technology, Charles Hull, formed a process of stereolithography (SLA) in the year of 1986, which had succession techniques, such as ink jetting, contour craft, fused deposition fabrication (FDM), and powder bed fusion. The manufacturing and logistics processes have the potential to be revolutionized thanks to the development of 3D printing, which is a process that has evolved over the years and involves a variety of methods, materials, and equipment. The construction, prototyping, and biomechanical industries are just some of the many fields that have found widespread use for AM. The popularity of 3D printing is very sluggish and restricted despite of its multiple benefits like reduced wastage, freedom of creation, and automation enhancement.

As new materials and AM techniques are continually developed, new applications are continually being discovered and developed. The expiration of earlier patents has enabled manufacturers to develop new 3D-printing devices, which is one of the primary factors contributing to the accessibility of this technology. This is due to the fact that the patents covering earlier versions of the technology have now expired. Recent innovations have led to a decrease in the cost of 3D printers, which has led to an expansion of the technology's applications in settings such as classrooms, homes, libraries, and laboratories. Because of the benefits of producing prototypes instantly and at a minimal cost, 3D printing was initially put to extensive use by architects and designers for the production of prototypes that were both aesthetically pleasing and functional. The incorporation of 3D printing into the product-development process has resulted in a significant reduction in the additional costs that are incurred during this stage. In spite of this, the utilization of 3D printing in a wide variety of activities, from prototyping to the production of finished goods, has only become widespread in the past few years. The association of high costs with end user's tailored and customized products present a challenge for manufacturers when it comes to the process of product personalization and customization. On the other hand, AM is able to print customized products in small quantities and at relatively low cost, making it ideal for use in 3D printing. This is particularly helpful in the biomedical field, which typically requires one-of-a-kind products that are customized for each individual patient. In 2020, Wohlers Associates envisioned that the production of 50 percentage of commercial products has now moved towards the additive manufacturing techniques. Further, customized products are trending nowadays, which have been fabricated by 3D printing [1]. Besides the commercial productions, the medical implants have garnered the attention of AM techniques in the medical industry for fabrication using patient-specific data (CT scan and MRI data) [2]. Hence, printing along three perpendicular axes has become very popular in the various areas of real life. Therefore, the construction field also turn towards suitable techniques of AM so that infrastructure development can be made by using 3D printing. Due to this, mass numbers of the houses were also made by WinSun in China using the layer fabrication technology with an effective cost price of $4800 USD [3].

As an alternative to conventional manufacturing methods, additive manufacturing has been adopted by modern industry due to various advantages in production systems. These advantages and benefits are attributed to the manufacturing of the complex components and intricate geometries with high rate of precision, less wastage of the material due to optimized fabrication technologies, infinite space for designing the component, and customization of the product as per requirements. Further, for the different techniques of the additive manufacturing, materials such as metals, ceramics, polymers, and their composites has also been proposed by several researchers. In the metal-based materials, researchers have explored aluminum, magnesium, steels, titanium, chromium-cobalt, and oxide dispersion-strengthened aluminum-based composite as an input material for metal 3D printing [4, 5]. At present, the aerospace industry is using the advanced metals and their alloys because of their uneconomical fabrication in the conventional techniques, more time consumption, and difficulties in tool handling, etc. Amongst several polymeric materials, polylactic acid (PLA) and acrylonitrile butadiene styrene (ABS) are most commonly polymeric material, which are used as a primary material in the process of extrusion-based 3D printing. In the category of ceramic materials, concrete is the conventional material being utilized for the construction of buildings using additive manufacturing. However, ceramics (such as hydroxyapatite, bio-glasses, tricalcium phosphates, zirconia) also have been used for the fabrication of bio-implants such as scaffoldings [6]. The anisotropic behavior of the printed parts discourages the substantial mechanical properties, ultimately, affecting the potential of additive manufacturing. As a result, it is vital to have an optimized pattern of additive manufacturing to reduce the deformities and flaws in the printed parts with a controlled anisotropic behavior [5]. Additive manufacturing has a tremendous proficiency for being used in wide ranges of the component geometries, shapes, and sizes ranging from macrolevel to microscale. Nevertheless, the precision of the printed parts is determined by the minimum part size (i.e., micro-sized to nano-sized) with respect to the accuracy. For an instance, the better resolution of the microscale components is very difficult in 3D printing, which further demand the post processing, such as heat treatments, for getting the good surface finish and enhanced layer bonding [7]. Furthermore, the limitation in the availability of the materials for the printing is posing a big challenge for introducing and implementing the techniques of AM for the societal applications. As a conclusive point, the demand for exploitation of suitable materials is also rising in parallel with the development of the AM techniques so that the improvement in the properties of printed parts can be done in relation to the intended applications.

The benefits of the three-dimensional printing technologies will go forward to emerge in society in due course within the research work. The research work taking place can help to predict the obstacles in the path of this technology to become standard. The association of CAD systems with AM enhances the capabilities of the present technique due to a user-friendly environment. It facilitates the advanced simulation attributed as a better design tool to access the life-cycle costs. These are some essential elements of AM that should be realized for better improvement. As an example of economic maintenance, the mold making and high-cost tooling has been replaced with direct fabrication using AM in the mass production. 3D printing is also associated with mass customization of series of the personalized products with unique designs while maintaining the economy. Such a kind of mass customization with conventional manufacturing systems is time consuming, which may acquire extra cost for post processing [8]. However, the resolution for the cost reduction must be done along with the improvement in the fabrication speed through the better design of the techniques.

The present chapter aims to provide a review of available additive manufacturing techniques, in terms of fundamentals, technologies, recent developments, and their application in several areas. In the different sections, the challenges in the development of various techniques have also been discussed to study the evolution of AM.

1.2 ADDITIVE MANUFACTURING TECHNOLOGIES

Additive manufacturing, in general, is described as a manufacturing technique in which the principle of layer addition is used for component fabrications. However, researchers have described AM based upon their techniques, such as deposition, curing, binding, and solidification. The part building occurs in several steps of layered curing or deposition. Always, the CAD model is an input to any kind of additive manufacturing technique, which is sliced further to generate the layered information

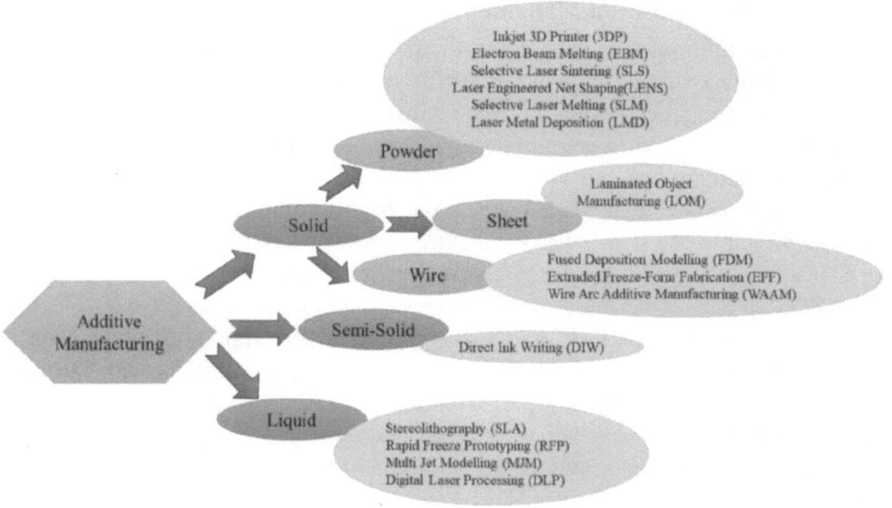

FIGURE 1.1 AM techniques based on material input.

of the components. The sliced part file is fed to the 3D printer for final fabrication of the component. In this way, the production process results in 3D printing with a fine layered resolution developed during the slicing process. With the growth and popularity of AM, several techniques of layered fabrication have evolved in modern industries with the ability to fabricate very large structures of intricate shapes with characteristics better than the conventional manufacturing process. These techniques can be further classified based on layer resolution, material input, part build size, fabrication speed, part accuracy, and working principles. The most popular available techniques are fused deposition methodology (FDM), selective laser sintering (SLS), SLA, selective laser melting (SLM), etc. However, in the present chapter, various AM techniques have been described based on the material used, as shown in Figure 1.1. A comprehensive study of these techniques can be done from the previous literature [9]. The recent development of AM, such as two-photon polymerization (TPP), electro hydrodynamic printing (EHP), non-contact nano- and microprinting, and projection microstereolithography (PμSLA) has also been discussed in the present chapter [10, 11].

1.2.1 Solid-Material-Based Additive Manufacturing

This section of the chapter includes techniques of additive manufacturing, which use the input material in the solid state. The input material can be in the form of solid wire (i.e., filament rolled over a spool), a thin sheet, particles and granules, etc. The most popular techniques in this category are fused deposition modeling (i.e., FDM), extruded freeze-from fabrication (EFF), laminated object manufacturing (LOM), inkjet 3D printer (3DP), electron-beam melting (EBM), SLS, laser-engineered net shaping (LENS), SLM, laser metal deposition (LMD), etc.

1.2.1.1 Fused Deposition Modeling

In FDM, a continuous filament of the thermoplastic material is deposited on the printing bed by passing through a nozzle after heating as shown in Figure 1.2. In the initial stage of the process, polymer filament is heated just below the liquidus temperature and then forced to extrude through the nozzle (i.e., having less diameter as compared to filament diameter) for depositing over the printing substrate. Thereafter, the subsequent layers are deposited one by one for the completion of the part. Hence, the basic principle of FDM is based on extrusion, which is also attributed as single-extruder as well as dual-extruder systems. For the polymeric material to be utilized in the FDM process, thermo-plasticity is a very important property due to which filament fuses and become harder after the printing to get a rigid permanent shape.

The fundamental process parameters of layer thickness, printing speed, extrusion width, infill angle and pattern, and part orientation influences the mechanical behavior of the component [12]. However, sometimes interlayer gap, warpage, elephant foot, and distortion are also identified as the limitation of the process, which also directly affect the mechanical properties [13]. The association of the simplicity, low cost, and

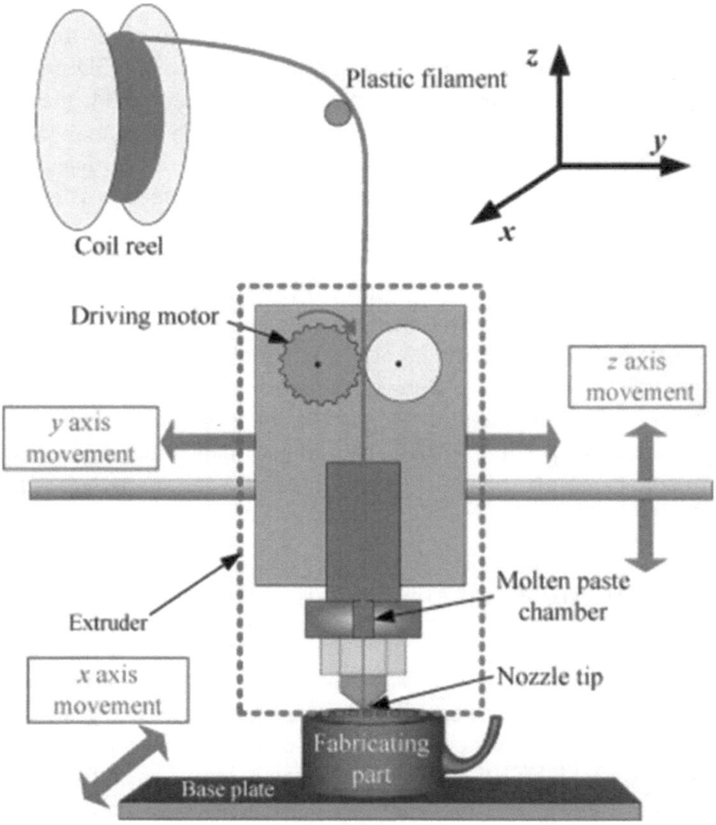

FIGURE 1.2 Fused deposition modeling [12].

low printing time make FDM advantageous. In the window of benefits, FDM also enclose some limitations of poor mechanical properties due to interlayer gaps, bad surface finish due to staircase effect, and limited material variety [12–14]. Further, thermoplastics can be reinforced with fibers as well as micro- and nanoparticles to develop the composite filament, which can be used in FDM to improve the mechanical characteristics of the printed components [15]. However, the control over the orientation, bonding between matrix and fiber, and vacancy at the interface is very limited, which is still a big challenge for the researchers [15, 16]. The capabilities of FDM has been increased tremendously with the development of composite filaments, which has led to several applications, such as sensing and shielding application, drug delivery devices, rapid tooling, aerospace and automotive applications, microfluidic devices and customized prosthetics, and orthosis [17].

1.2.1.2 Extruded Freeze-Form Fabrication

Extruded freeze-form fabrication is the technique of 3D printing in which most of the functionally graded materials, like ceramics, can be used. The aqueous mixture of colloidal slurry is taken as the input material in EFF. The similarity of this technique with other extrusion-based printers is that the input ceramic material is developed by adding suitable organic binders [18, 19].

However, the percentage of the binder in the present technique is only 2–4% with percentage loading of the solid ceramic particles ranging from 40–50% or more. As the name given freeze form, EFF fabricates the parts below the freezing point of the water and develop a green component. Below the freezing point, fabrication enables very fast solidification, which make the process feasible. In comparison the robocasting, part size fabricated in EFF is larger. The process does not consist of any harmful chemicals, making it the safest technique. The water is self-acting as a media of dissolution as well as a binder for the particles [20]. The schematic of the extruders used in the EFF process is shown in Figure 1.3. The EFF process was developed for the fabrication of various functionally graded materials and ceramics to develop several composite structures. It facilitates the high ceramic material loading, which further help to maintain the shape of the component. Some featured components in the aerospace industry has been fabricated by the EFF process like nozzle throat, nose cones of missiles, and hypersonic vehicle components for spacecraft propulsion, which are associated with high performance [18].

1.2.1.3 Laminated Object Manufacturing

Laminated object manufacturing is the technique of additive manufacturing in which lamination of the input material fabricate the parts after a successful cutting of the extra material. The raw material is taken as the sheets of paper, polymers, and thin metals, which are stacked one layer over another layer of the same material during the process. The basic phenomenon of the LOM is shown in Figure 1.4.

Along with layer-by-layer lamination, layer-by-layer cutting is also followed for defining the contour of final component. The cutting of the laminated sheets can be performed either by mechanical cutter or laser. The working of the techniques is based on two methods as form-then-bond and bond-then-form. The bond-then-form

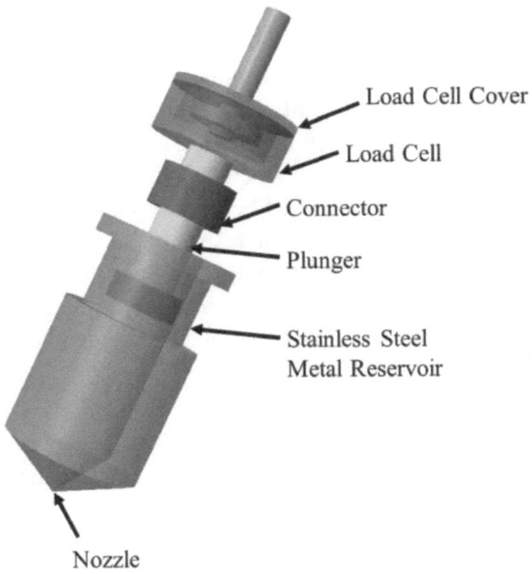

FIGURE 1.3 Extruded freeze-form fabrication process [20].

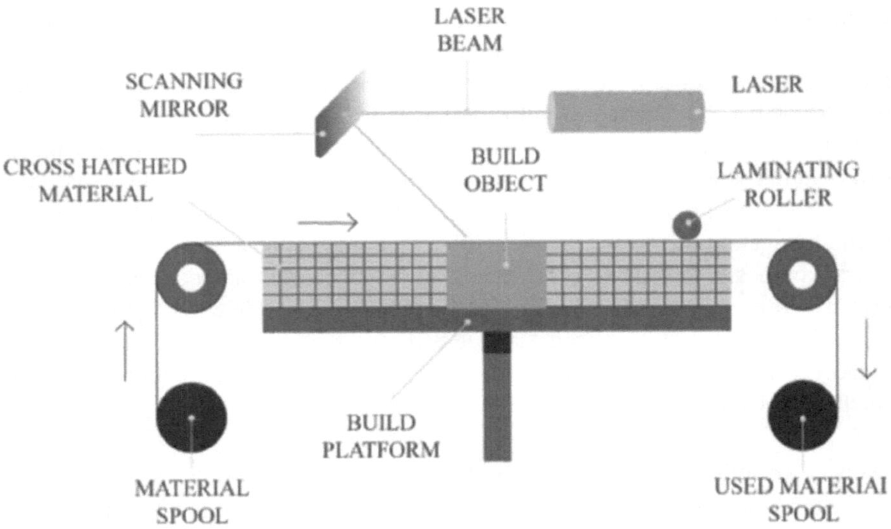

FIGURE 1.4 Techniques of laminated object manufacturing [21].

methodology is effectively used as it is economical, less time consuming, and suitable for metallic sheets [22]. Even, ceramic material can also be employed in this process in the form of sheet for high-strength components. In most of the instances, the process requires post processing, such as heat treatment, which depends upon the material as well as properties required. The process also has been developed using ultrasonic additive manufacturing in which ultrasonic vibration are used to laminate the sheet using the process of welding [23]. For the fast prototyping, LOM can offer the low cost with large size structure. However, it is associated with drawbacks of poor surface finish, bad dimensional accuracy, and time-consuming post processing. The low cost and large size feasibility has facilitated the LOM process to be widely used for the prototype fabrication and some architectural components. Along with these, the LOM process also has its application in rapid tooling, ceramic processing, sand casting, and investment casting, which can help to reduce the cycle time of the process [24].

1.2.1.4 Selective Laser Sintering

SLS, developed in 1990, works on the principle of sintering of the powder material in a selective manner [25]. It is a kind of powder bed fusion process, which consists of spreading the powder material on the substrate in the form of a thin layer. In this laser beam is used to heat up the particles so that they can join by sintering. After the successful completion of one layer, the re-coater spreads the powder over the prebuild layer and rolled over it to maintain the uniform thickness of the layer as shown in Figure 1.5.

In the post processing, a compressed air or vacuum is used to remove the loose powder for making the part suitable for further processing. In this technique, the main powder material is also acting as a support material to develop any kind of intricate shape. The advantage of this techniques is that we can get the component with a density close to the solid component [27]. SLS is preferable for polymeric powders and low melting point metals and alloys. The better distribution of particles of the low melting point material promotes the sintering process under the effect of laser [28]. By looking at all these features of SLS, it has been applied to various areas

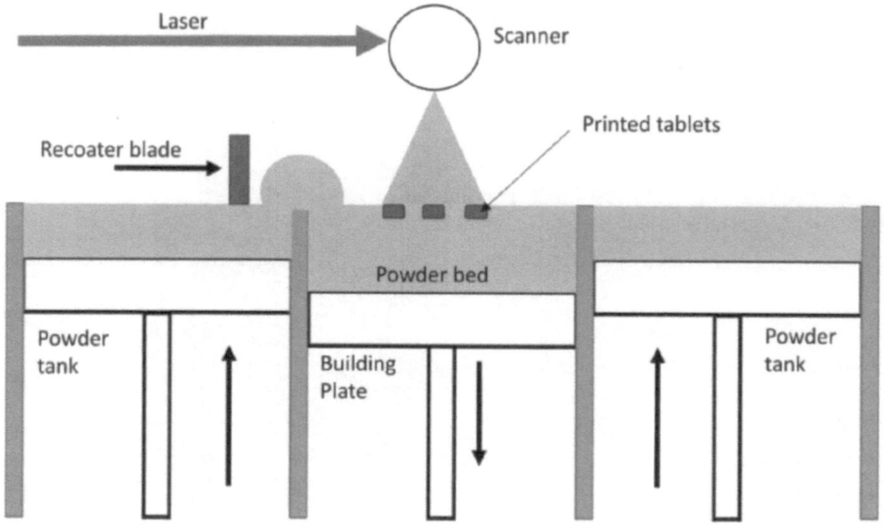

FIGURE 1.5 Schematic of selective laser sintering [26].

like the aerospace industry for high-strength components, the automotive industry for the weight reduction of the parts, the wind tunnel for good dimensional stability, rapid tooling, injection molding, biomedical implants, hearing aid components, microelectronics components, and medical devices [29].

1.2.1.5 Selective Laser Melting

In selective laser melting, part is fabricated in a layer-by-layer fashion from the fusion of the powder particles under the effect of heat produced via laser light. However, in comparison to SLS, the principle of the melting is used to join the particles of the powdered raw material. Hence, high melting point material like metal powder can be employed in SLM. During the process, powder material is melted using a thermal source like a laser and the solid part is obtained after the solidification of the melted powder particles [30]. Similar to SLS, residual and unused powder particles are cleaned from the components [31, 32]. The working principle of the SLM process is clearly explained in Figure 1.6. SLM is generally used for the fabrication of the components from the metallic materials like aluminium, titanium, nickel alloy, and iron-based alloys, etc. [33, 34]. The oxidation process during the fabrication is avoided by providing the inert atmosphere using nitrogen or argon gas [34]. The advantage of the SLM process is that we can fabricate near net shape components with less requirement of post processing, high particle density, thermal stability, and better mechanical properties. Along with additive manufacturing, the SLM process has also been adopted for the repair work of the damaged or eroded components. In the repair work, high-power laser Nd:YAG is used for melting the material and depositing over the cracks [35]. Presently, SLM is being widely used for the development of medical implants, such as bone tissue scaffold, dental implants, hip-joint, and many other orthopaedic implants. Furthermore, the SLM process has been applied in various areas like light weight structures, heat exchangers, wave guide filter, aerospace and automotive industry [34].

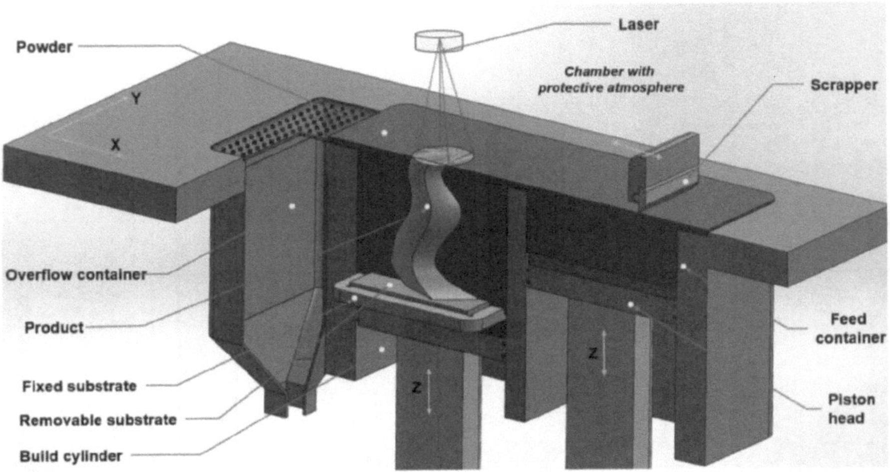

FIGURE 1.6 Working principle of selective laser melting.

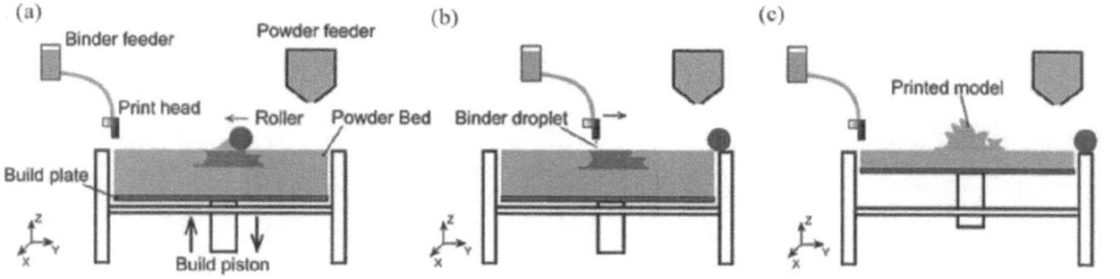

FIGURE 1.7 Working of three-dimensional printing (a) powder distribution, (b) binder spreading, and (c) part completion [38].

1.2.1.6 Three-Dimensional Printing

The powder material based 3DP involves the usage of liquid binder as a component of the process, which help to selectively join the particle for fabrication purpose. 3DP depends upon the process parameters of rheology and chemical composition of binder, size, and morphology of powder particles, binder spreading speed, chemical affinity between the binder and particles, and requirements of further processing [13, 28]. The most commonly used material is stainless steel powder, which fabricates the part after the spraying of a suitable binder during the process. In this process, powder material is loaded into a powder feeder controlled by a piston arrangement as shown in Figure 1.7. The piston helps to feed the material on the substrate after a successful completion of the previous layer. The technique is efficient to fabricate the overhang structure without the requirement for a support material as the unbound powder helps to make the unsupported structures. However, the technique is also associated with the disadvantages of limited materials, high porosity, and high cost with high time consumption. The general application of powder-based 3DP is to fabricate the dies for injection molding and rapid tooling [36, 37].

1.2.1.7 Electron-Beam Melting

Electron-beam melting is the additive manufacturing process, which uses a high-energy electron beam as a source of energy for making the full density components using the melting principle [39]. The benefits of using electron beam in spite of the laser is that one can get increased penetration depth with elevated velocities of the scanning by which the build rate of the process increases.

The basic principle of EBM is that a focused electron beam is deflected in such to solidify the selected powder material over the substrate ranging in the layer thickness of 20 to 100 μm [40]. All the process is conducted in a vacuum with a small amount of helium gas flow to avoid any charge build up over the material and maintain thermal stability as shown in Figure 1.8. After the completion of the process, helium gas pressure is increased for fast cooling of the fabricated components. The fabricated part is further post processed to remove the adhered agglomerated powder particles be means of sand blasting operation. As the operation is performed inside a control environment, therefore, the obtained powder can be recycled for reuse [39, 41]. EBM has better mechanical properties and highly dense parts without

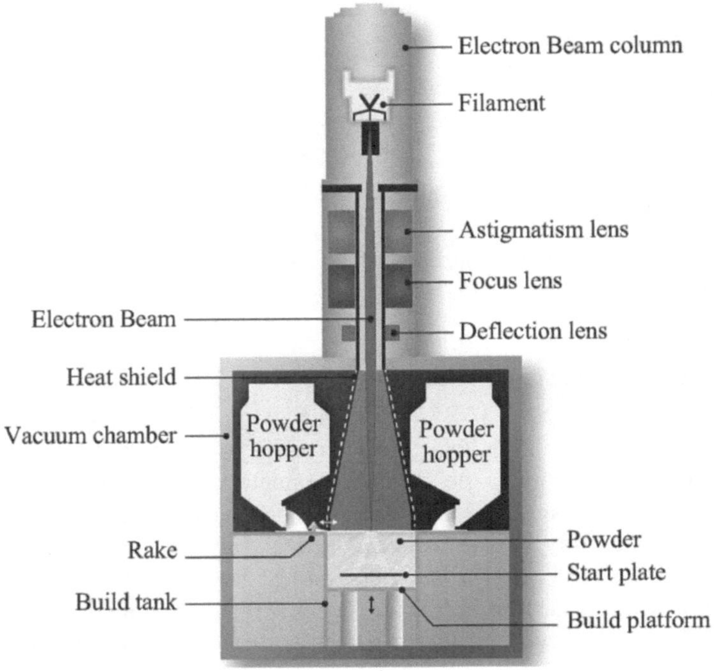

FIGURE 1.8 Schematic of electron-beam melting [39].

any alteration in the composition. Besides these advantages, the discrepancies are also associated with the process in terms of lower stability due to charged particles, melt ball formation, inadequate energy density transmission and delamination. Various fields have been identified in which EBM can be successfully employed for applications, like medical components, injection molds, bearings, artificial joints, dental prosthesis, high duty parts, cutting tools, aerospace nozzles, turbine blades, and cellular materials [42].

1.2.1.8 Laser Engineered Net Shaping

Laser-engineered net shaping, evolved in 1970, is a technique of solid free form fabrication in which direct metal deposition techniques is used to fabricate the metallic components. The process is characterized to use the metallic powder with size of 150 µm. In this process, the fabrication is done by injecting the metal powder in the melt pool as shown in Figure 1.9.

The carrier gases are used to inflow the metal powder, which generates the metal melt pool by the using high energy laser beam (e.g., Nd:YAG) [44]. The subsequent addition of powder material and generation of melt pool results in the fabrication of the final component in the layered manner. In this process, a metallurgical bonding of the particles is observed, which provides a high dense part. Similar to EBM, this process is also conducted in an inert atmosphere, hence, powder material can be recycled [44]. The developed process is so revolutionized that it can fabricate a complexed

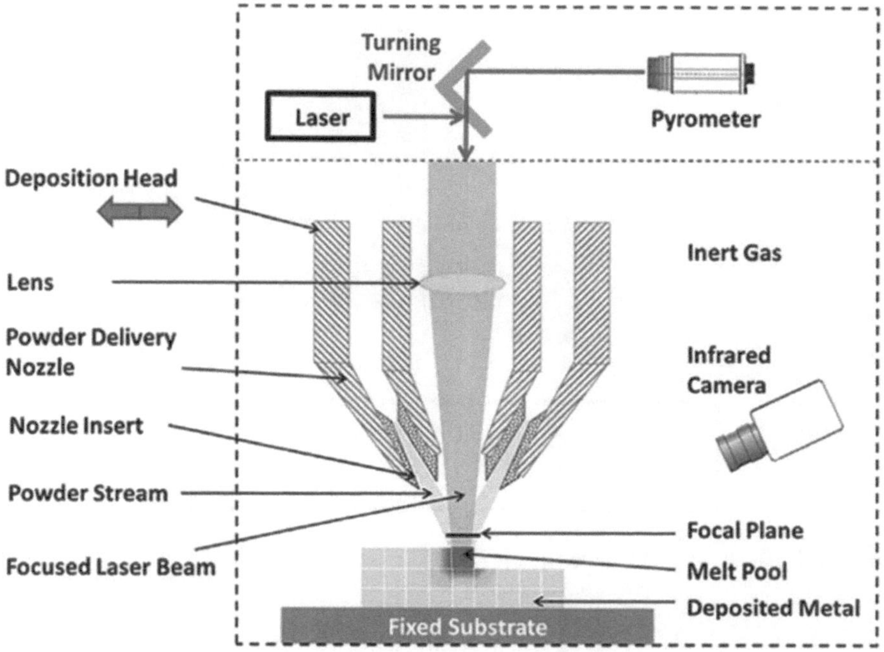

FIGURE 1.9 Laser-engineered net shaping [43].

shape, metal tooling, and real-life components (small lots) without any limitation. However, it is associated with the limitations of uneven temperature gradient due to variable eating. Due to this higher residual stresses occurs in the components [43]. The metal powder such as H13 tool steel, titanium, 316 stainless steel, and nickel-based alloys can be used to fabricate functionally graded components. The availability of various metal powders has facilitated the LENS process to be used in complex tool making, aerospace and automotive components, medical device manufacturing, surgical instruments, and injection molds with conformal cooling [45].

1.2.1.9 Laser Metal Deposition

LMD technique uses a high-power source of energy to fabricate a component by processing of powder material as shown in Figure 1.10. Laser is used as a high energy source that provides a heat energy to metal powder, which thereafter deposited over a platform for the fabrication process [46]. The powder material is delivered at the spot of laser focus by using any inert gas, which is further melted and solidified in the different layers. LMD process is widely used for the fabrication of near net shape components. Further applications of LMD processes are laser cladding, repairing of damaged parts and fabrication of functionally graded components. In comparison to the SLM process, LMD have more benefits of inexpensiveness, very narrow heat-affected zone, fabrication of unmachinable material components, less distortion zones, and being environmental friendly [47]. Because of less thermal gradients,

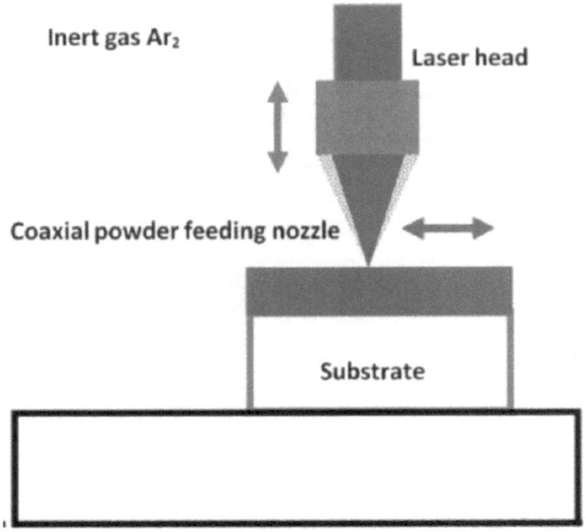

FIGURE 1.10 Laser metal deposition [47].

heat-affected zone are very less, which results in negligible residual stresses. In comparison to conventional manufacturing, complex structure fabrication, low material wastage, less production cost are the key features of LMD. LMD is known to be a powerful process, which is still finding new applications in the areas of repairing components (tooling surfaces, engine components, turbine blades, etc.), medical implants, prototyping, and surface cladding [48].

1.2.1.10 Wire Arc Additive Manufacturing (WAAM)

WAAM is a process of material fusion consisting of melting the metal electrode using the heat energy obtained from the arc produced. The material fused because of this available heat energy is deposited over the printing substrate in the layered form to develop a complex structure. In these techniques, material addition is also related to the deposition of the weld bead, which resembles the electric arc welding as well as cladding process. The process is attributed to the capacity of manufacturing the large structures, elevated deposition rate, capable of holding different arc generating sources along with their alignments and movements [49]. A wide range of the materials, such as aluminum, titanium, steel, nickel alloys, etc. can be fabricated with a fabrication time 60–70% of the conventional manufacturing processes. The heat source in the WAAM process has been used for their classification, which are based on (i) gas metal arc welding (GMAW), (ii) gas tungsten arc welding (GTAW), and (iii) plasma arc welding (PAW) [50] as shown in Figure 1.11.

The production rate needed and processing condition helps to select the suitable WAAM technique. Based on the characteristics, the WAAM process has been applied to the several areas like defense equipment, aerostructure components, functionally graded material fabrication, disks and blades of gas turbines, and near-net shape manufacturing [51].

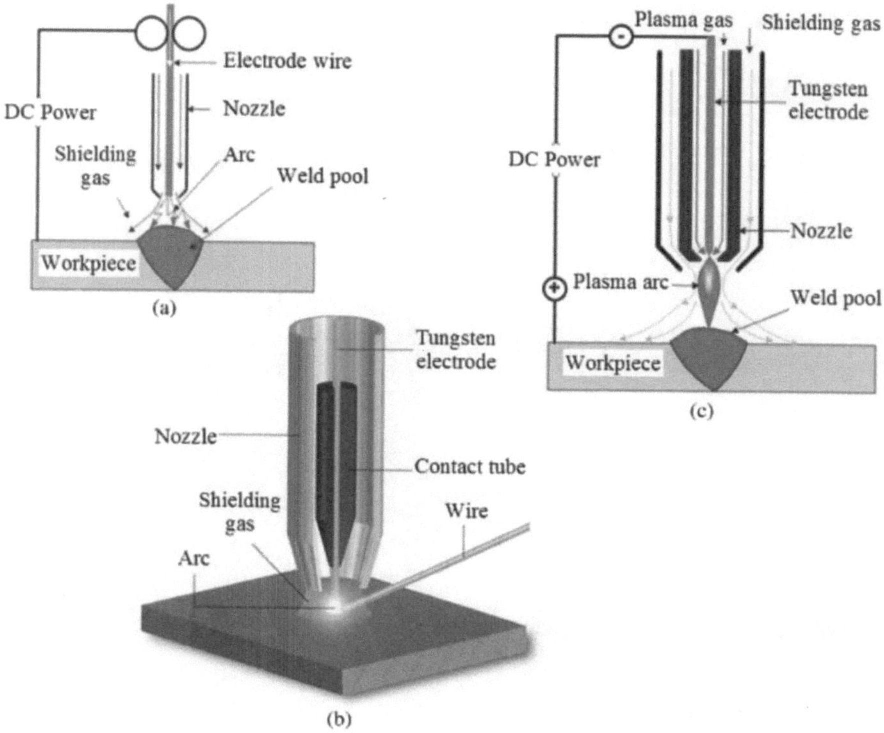

FIGURE 1.11 (a) MIG, (b) TIG, (c) plasma arc welding principle of WAAM [49].

1.2.2 Semi-Solid-Material-Based Additive Manufacturing

This section of the chapter includes the techniques of the additive manufacturing, which are using the input material in the semi-solid or mushy state. Semi-solid material is fabricated by dissolving the polymeric materials in the suitable solvents, which provide a paste-like formation of the material.

1.2.2.1 Direct Ink Writing (DIW)

Direct ink writing (DIW) works on the principle of the extrusion in which as-prepared material solution is extruded through a dispensing tip and deposited on the printing bed in the form of stacked layers. It is also named as robot-assisted deposition (RAD) and direct write fabrication (DWF) [52, 53]. The technique uses the non-Newtonian fluid in the form of viscous slurry at the room temperature. Hence, the rheological properties are very important to control while preparing the material. The prepared material is loaded into the plunger system attached to the print head and extruded by the help of force applied in either mechanical or pneumatic way as shown in Figure 1.12 [54]. Hence, by the synchronized motion of the robotic-arm (print-head) and printing bed, the fabrication of the component is accomplished. Depending upon the material, the post processing may be required for getting the better mechanical

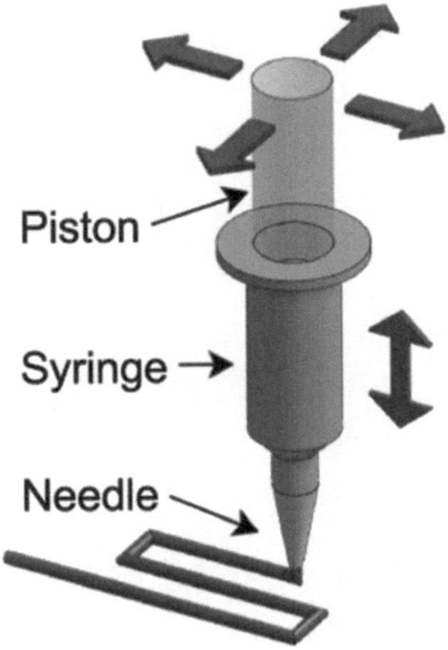

FIGURE 1.12 Direct ink writing additive manufacturing [57].

properties, removal of binder agents, and densification of the part [55, 56]. DIW additive manufacturing is a technique with a versatility to fabricate the parts of various industries, such as biomedical, electronics, constructions, and composites. Hence, it allows the variety of the materials, such as polymers, metals, ceramics, and composites. Recently, DIW has also found its application in the optical region by developing a water sensitive glass.

1.2.2.2 Solvent Cast 3D Printing

Solvent cast 3D printing (SC3P) is a technique of fabrication in which the material is prepared by the help of a suitable volatile solvent material. The fabrication of the part is accomplished after the evaporation of the solvent from the deposited structure as shown in Figure 1.13 [58]. The rigidness of the printed component is dependent upon the efficiency of the evaporation as well as the extrusion diameter of the nozzle. In this technique, the material ink, which is having a semi-solid structure, is extruded through the nozzle by the application of the pressure by means of air, screw, and plunger, etc. Finally, the shear flow inside the printing nozzle is relaxed while coming outside and deposition occur on the substrate. The prepared mushy state materials are termed as printing ink, which are generally prepared by using polymeric materials, like polycaprolactone (PCL), polylactic acid (PLA), poly lactic co-glycolic acid (PLGA), polyvinylidene fluoride (PVDF), etc. The available evaporative solvent solutions are dichloromethane, ethanol, acetone, dimethylformamide, etc. [59–61]. With a suitable mixture of polymers in solvents, SC3P has been widely used for cardiovascular application, bone tissue scaffold, metallic structures, and multifunctional microsystems [62–64].

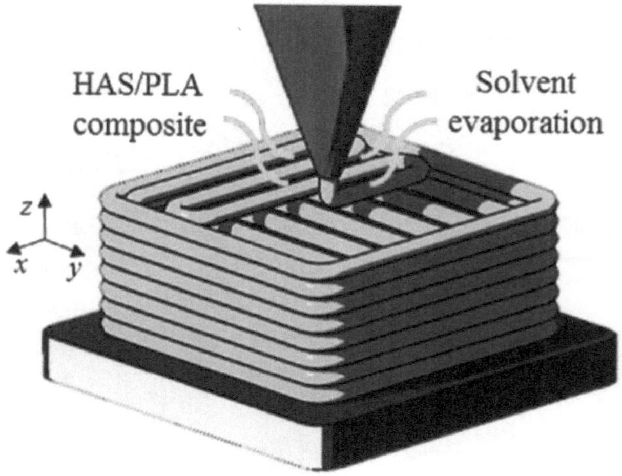

FIGURE 1.13 Solvent cast 3D printing [58].

1.2.3 Liquid-Material-Based Additive Manufacturing

The techniques, in which the input material is used in liquid state, are classified in the category of liquid-material-based additive manufacturing. In these techniques, the rigid structure from the liquid material is generally fabricated by using the process of polymerization, solvent evaporation, UV-curing, and rapid freezing, etc. SLA, digital light processing (DLP), multi-jet modeling (MJM), and rapid freeze prototyping (RFP) are the various techniques that are using the input material in the liquid form.

1.2.3.1 Stereolithography

In 1986, the techniques of component fabrication by using a photosensitive resin was termed as SLA. SLA has been evolved as a technique of fabrication with good surface finish, high accuracy, and faster production [65]. During SLA, physical components are fabricated by using photosensitive resins, which undergo the process of photo-polymerization under the presence of light (i.e., UV light) and form a solid polymeric structure as shown in Figure 1.14.

After the printing, there may be any requirement of post processing, such as photo curing and heating for some of the components to get the desired rigidity, which are also essential for the completion of crosslinking. During the printing, the main material itself acts as a support material for developing any overhang or cavity features [67]. Highly detailed parts with microfeatures and dimensional accuracy can be easily fabricated via SLA [66]. Among the several advantages, it has some disadvantages like brittleness, low impact resistance, and loss of mechanical properties with the exposure of the light for a long time. SLA has the widest area of applications, like automotive, medicine, bio-implants, treatment of diseases and surgical preplanning, etc. [27].

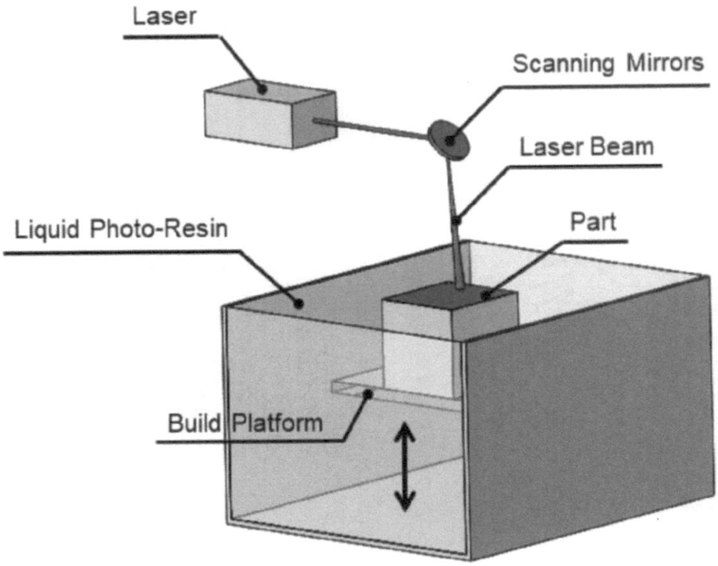

FIGURE 1.14 Schematic of the stereolithography process [66].

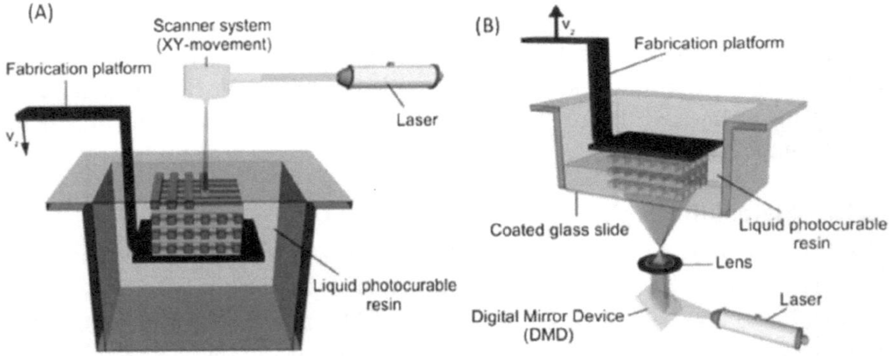

FIGURE 1.15 Comparison between (a) SLA and (b) DLP.

1.2.3.2 Digital Light Processing

The DLP technique of image projection technology, similar to SLA in which photosensitive resin is cured in the presence of light and part, is fabricated. However, light source is a conventional light source (such as arc lamp or projector light) with non-coherent properties [68]. In spite of scanning on a single spot as in SLA, DLP scan the complete single layer in a single step as shown in Figure 1.15. Due to this, the speed of process is more than SLA.

The process is highly influenced by the value of layer thickness and time of exposure. On the other hand, part accuracy is more in SLA in comparison to DLP.

Therefore, the time consumption in DLP is more than SLA. The digital mirror device in DLP helps to adjust the projection planes, which ultimately tends to change the resolution of the part. Thus, the DLP printing has a comparatively superior resolution, which is very near to SLA [69]. The literature has found that many of the resins are not compatible with the living tissues and cells. Hence, application of DLP in the biomedical field is based on only some limited materials. In general, application of DLP includes dental application, jewelry applications, entertainment industry models, and hearing aids [69].

1.2.3.3 Multi Jet Modeling

In the multijet modeling technique, the print head has the numbers of the dispensing system, which selectively deposits the material on the substrate. The material is deposited in the form of the droplet, which are comprised of the photosensitive material resin mixed with secondary type of the materials (like binders) [70]. The secondary material is sometimes used as a support material during the fabrication process. When the material is jetted over the substrate, the UV lamp gets switched on to cure and solidify the deposited material. In the final stage of every layer, a leveller planes the upper surface to produce a smooth and flat layer as shown in Figure 1.16. The technique uses a multimaterial approach by using multinozzle dispensing system. Hence, composite material fabrication can be done without separate material preparation as reinforcement or blending. For

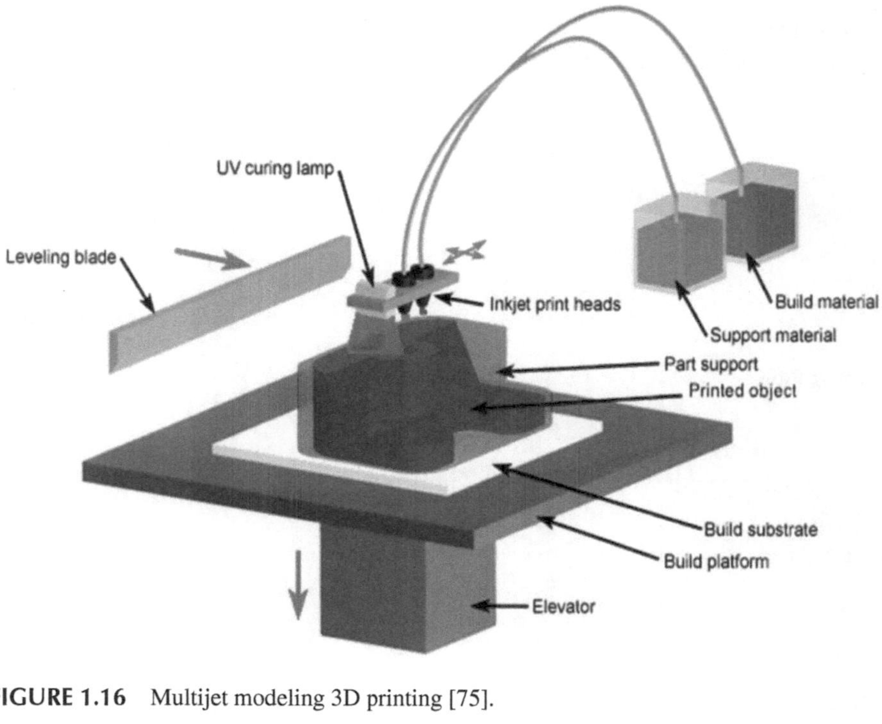

FIGURE 1.16 Multijet modeling 3D printing [75].

a better surface finish and cleaning of the part after printing, post processing is required, which also helps to remove the support material [70, 71]. The lesser time required, cost effectiveness, high surface finish, better quality, multicolor printing and safety are the salient feature of this process, which make it suitable for use in fast prototyping. Besides these advantages, the process has the limitations of material selection, build processing, reduced strength, and bad surface quality at the bottom [72, 36]. In lieu of these benefits, MJM has applications in microchannel fabrication, molds for investment casting, surgical tools, dental prosthesis, and veterinary medicine [73, 74].

1.2.3.4 Rapid Freeze Prototyping

Rapid freeze prototyping is a sold free form process of 3D printing, which produce the part by freezing the material in a layered fashion. The selective deposition and rapid freezing of the material (water or brine) results in the fabrication of physical components. The process consists of two-way deposition of the material either in continuous form or dropwise manner. The temperature of the printing area is maintained below the freezing point of the material. Generally, water is used as a material along with brine as a support material [76]. The material is extruded through the nozzle in any way (i.e., continuous form or dropwise manner) and get frozen after the deposition over the substrate as shown in Figure 1.17. Due to low temperature (approx.–20°*C*), frozen material stick to the substrate, which helps to deposit the subsequent layers [77]. In the continuous mode of fabrication, the ratio of flow rate to printing speed is very much crucial, which can affect the quality of the part, as shown in Figure 1.18.

Once, the printing is done the support structure is removed by dissolving the component in a suitable solvent below the freezing point of the main material. The solvent

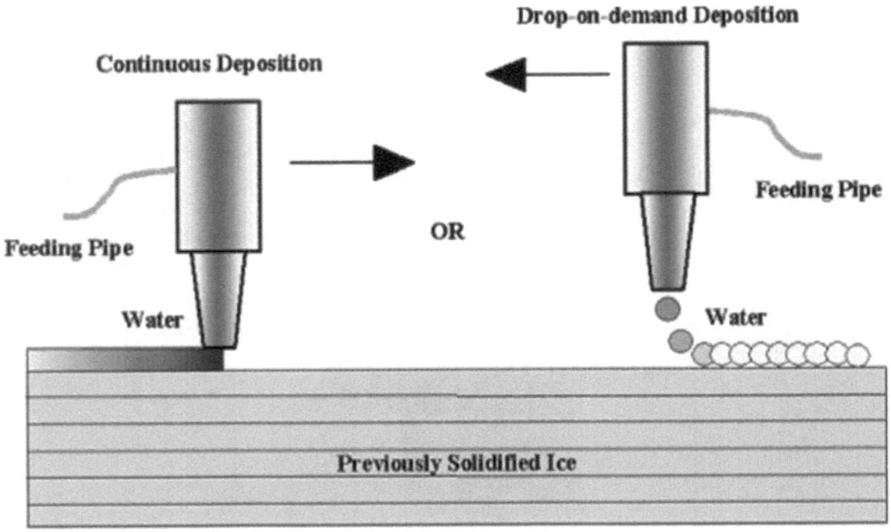

FIGURE 1.17 Schematic of rapid freeze prototyping process [76, 77].

FIGURE 1.18 Effect of ratio of flow rate to printing speed on the material deposition [77].

used for support removal should be immiscible in the water and have different density [76, 78].

1.3 RECENT DEVELOPED ADDITIVE MANUFACTURING TECHNIQUES

The discovery of additive manufacturing has made a new domain in the manufacturing field by which interdisciplinary work has grown between science and engineering with optimum results. The capability of producing new opportunities for the development of products and services define its real power that can effectively change the world in a significant way. AM also proved itself as a revolutionized technology for the development of components for the marine, aerospace, automobile, and defence industries. Thus, such kinds of revolutions have also given the momentum to AM for developing the novel kinds of techniques, which can support the modern manufacturing industries with a considerable amount of economy. In this regard, researchers have worked on the development of some advanced methods of AM with tailorable parameters, which can develop the customized product. Hence, in this section, several newly developed techniques of AM have been discussed, which are growing very fast in the field of advance manufacturing [10, 11, 79].

1.3.1 Two-Photon Polymerization (2PP)

Two-photon polymerization is a technique for fabrication of a nanoscale component by the absorption of two photons and polymerization of the photoresist resin material [80]. In this process, two photons at longer wavelength near to the range of infrared spectral region are used because of their higher spatial resolution, lesser optical scattering, and better penetration depth. [81]. The spatial resolution of two photons helps in the fabrication of physical structures with sub-micro- and sub-nanoscales, which is also attributed to nonlinear behavior of photons' absorption phenomenon. This spatial confinement restrict the polymerization reaction to occur at the focal volume of high-intensity laser as shown in Figure 1.19. Further, nonlinear behavior and focal point polymerization tend to produce a structure with very fine resolution [82]. Hence, a liquid material is converted into a solid structure by the process of polymerization using two photons. The 2PP process has the advantages of negligible linear absorption causing higher penetration and quadratic behavior of

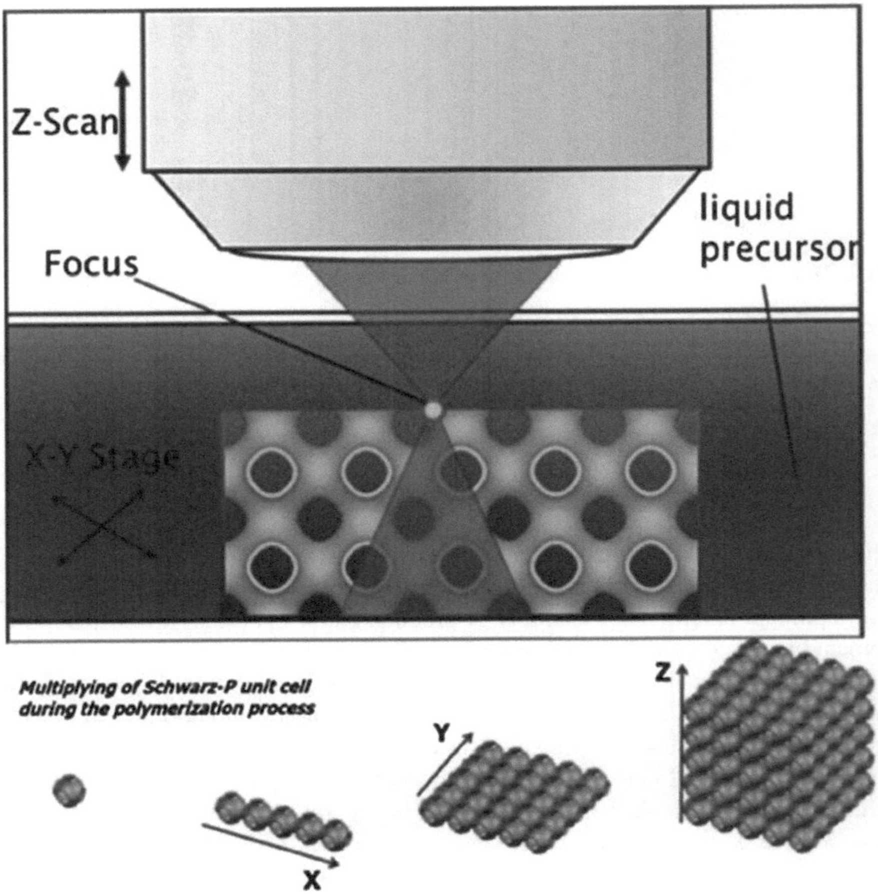

FIGURE 1.19 Schematic of two-photon polymerization [84].

polymerization process on the laser intensity causing better spatial 3D resolution for accuracy [83]. Hence, the unique features of 2PP allow fabrication of a complex and intricate shapes in sub-micrometer range. Hence, 2PP has its application in several application, like microneedles, microfluidic channels, microbiotics, bio-sensing, and micro-optics, etc.

1.3.2 Electrohydrodynamic Printing (EHP)

Electrohydrodynamic printing (EHP) is an additive manufacturing technique with high resolution in which the limitations of conventional inkjet printing has been eliminated by the use of highly precise micro-sized nozzles [85, 86]. The printing process in EHP is based upon the electric field, which helps to withdraw the ink material for printing. The electric field results in the mobile ions within the polarizable fluid, which are allowed to gather on the surface on the substrate. Due to this, electrostatic Coulomb force produces results in the development of a meniscus near

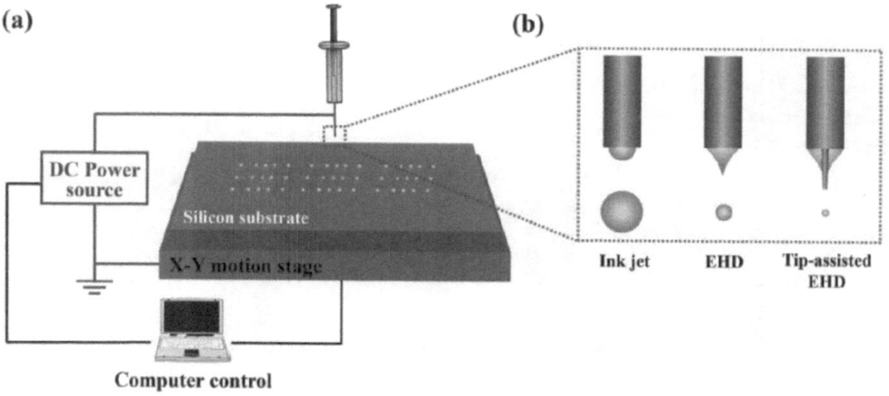

FIGURE 1.20 (a) Schematic of electrohydrodynamic printing, (b) nozzle configuration with substrate [85].

the end of conical shape tip. As the electric field is increased to a large extent, the surface tension also increases at the apex of the cone due to charge repulsion near the surface [87]. Hence, a jet or droplet is deposited over the printing substrate, which is significantly smaller than the size of the nozzle tip diameter as shown in Figure 1.20. Thus, micro- as well as nanoscale features can be fabricated with high resolution.

Based on the process configurations, EHP can be performed in different modes like electrospinning, electro jetting, and electro-spraying, which are generally affected by input raw materials [88, 89]. On the other hand, the input material used for EHP are similar to the materials, which are being used for the electrospinning process. The EHP process has been applied for several fields like MEMS, bio-scaffolds, conductive material pattern printing with greater than 10 μm, and graphic arts. It is envisioned that the proposed hybrid EHD printing technique might provide a promising strategy to fabricate multifunctional micro/nanofibrous scaffolds with biomimetic architectures, electrical conductivity and even bio-sensing properties for the regeneration of electroactive tissues [90].

1.3.3 Projection Micro Stereolithography (PμSLA)

The recent development in stereolithography with a high throughput, which is capable of fabricating the complex 3D geometries with micrometer size of the range of 1 μm is famous in the name of projection-based stereolithography (PμSLA) [91]. Both, SLA and PμSLA use the photopolymerization process in which liquid photosensitive monomers combine together in the presence of light to form a polymeric solid structure as shown in Figure 1.21.

However, the novelty in the PμSLA technique is that this technique projects ultraviolet light through a resistive mask for polymerization of the entire layer's cross section in a single scanning. On the other hand, the available SLA technique does selective scanning, which polymerize the resin onto a small spot falling under the light. The micrometer resolution of PμSLA depends upon a digital mask having multiple

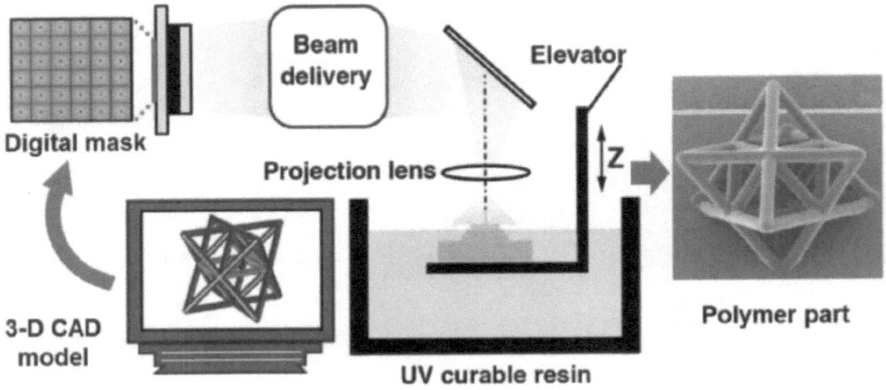

FIGURE 1.21 Schematic fabrication of component with PμSLA [10].

micromirrors. A 3D object is developed when UV light scans a digital image, which is projected through a digital mask on the upper layer of photosensitive material. For the digital mask, liquid crystal display (LCD) has been used as a dynamic mask by the researchers for easily obtaining the detailed pattern of each layer [92]. However, resolution is very limited as high pixel size in the usage of LCD. The researchers have also developed a digital masked device having a large number of micromirrors, which is capable of producing the parts lower than 1 μm. In general, numbers and the size of mirrors define the quality of the object fabricated. A large area PμSLA has also been developed with a combination of conventional SLA with digital light processing (DLP), which is capable of fabricating a multiscale 3D component. The resolution of 2PP is much higher than PμSL. Still, PμSL has several benefits compared to conventional SLA, like higher resolution, highly functional materials, shape memory polymer, and metallic components fabrication with polymer blending [93].

1.4 APPLICATIONS

1.4.1 Manufacturing Industry

The product development consists of one of the important steps of prototyping, which is the fundamental feature of additive manufacturing. In the product life cycle, fabrication of components, at a different scale than the real product, is started once the conceptualization is completed. This scaled product in called a prototype, which is very much useful in the analysis of the part and helpful in communicating main ideas of the product. Additive manufacturing is a kind of direct manufacturing, which eliminates the extra tooling like cutting tools, molds, dies, etc. Further, it also reduces the fabrication steps, like assembly, which is prominent in conventional manufacturing. Due to this, fabrication time gets reduced in AM with a higher rate of production. Nowadays, multimaterial and composite material additive manufacturing techniques are growing very fast, which are very helpful in tailoring the properties of the components [94].

1.4.2 Aerospace

In the aerospace industry, the components have complex and intricate geometries, which pose some limitations in conventional processes for the fabrication. The material used for fabricating such components are titanium alloys, nickel superalloys, special steels, or ultrahigh-temperature ceramics, which are difficult, costly, and time-consuming to manufacture. Therefore, the AM techniques are highly suitable for the aerospace industry. In aerospace, both non-metallic as well as metallic components are being developed using several AM techniques. Direct energy deposition (DED) has been used for fabrication of metallic components while SLA, FDM, etc. for non-metallic components [72]. Likewise, titanium structural components were fabricated by plasma deposition rapid technique by Norsk Titanium AS for Boeing 787 Dreamliner [95]. Araine 6 nozzle (SWAN) was firstly developed by Airbus Safran Launchers for Vulacn 2.2 by employing DED [96]. For high-precision parts, PBF techniques have also been employed. General Electric Aviation industry manufacture a jet engine component with advanced design in which an intricate shape for better cooling has been developed [97]. PBF helped in the improvement service life five times with less numbers of parts. Further, non-metallic components have been tried by NASA aeronautics for gas turbine engines. The components were fabricated using pure polymers, polymeric material as matrix and composite materials in FDM, binder jet AM, etc. [98, 99].

1.4.3 Biomedical Industry

The application of AM techniques has also found its way into the biomedical industry for developing, bio-material, bio-implants, drugs, and bio-medicines, etc. This broadening space of AM has facilitated the medical practitioners to develop tissue scaffolds, bone tissue implants, medical 3D-printed medicines, medical device artificial organs, live cell incorporation in tissue constructs, and biological chips [100]. In the tissue engineering, scaffolding plays a vital role in the form of extra cellular matrix for the interchanging of the cell while repairing the bones. Thus, the AM technique assisted effectively for the direct fabrication of tissue engineering scaffold using biocompatible and biodegradable polymers [101, 59, 60]. Among the techniques, SLS, 3DP, FDM, and SLA has been extensively used for such applications [102–105].

Dental implants have also identified AM as the potential techniques for the development of complex structure of human dentures. EBM and SLM techniques have been utilized for making the dental implant from the patient-specific data as obtained from CT scan and MRI data. Moreover, other implants such as acetabular cups, shoulders, knees, hips, and spinal have also effectively fitted the patient anatomy after using AM [106, 107].

1.4.4 Electronic Industry

Multi-layered PCB circuits has also used additive manufacturing for the fabrication with a reduced lead time from design to prototype fabrication. PCB fabrication is done by direct ink writing principle in which a viscous ink is deposited on the substrate. The ink is developed by a highly conductive material, which is also helpful in printing of double-sided chips. The silver nanoparticles viscous ink is used for the PCB fabrication because of their high conductivity and mechanical reliability. With a micro-sized nozzle printing, 3D-printed PCBs satisfies the purpose of conventional PCBs as well as modern integrated circuits [108].

1.4.5 Food Industry

Food printing with customized images, flavor, and texture is getting popular in the modern food industry. The food printing process does not degrade any nutritional value and allow one to add the more nutritional ingredient in a particular food during the fabrication. Hence, the foods can be printed in relation to personal needs and individual tastes. The printing material, which have been used in food printing industries, involves non-printable traditional food materials, native eatable printing materials, and alternative ingredients for nutritional values. The recipes in the food printing are classified as traditional recipes and elemental-based recipes, which are highly effective for the customization. The areas in the food industries where AM techniques have been applied successfully are the sweet industry, beverages, military foods, space food, etc. [109].

1.4.6 Construction Industry

The construction industry has also gotten the attention of additive manufacturing with an aim to improve the infrastructure construction, labor cost reduction, worker safety, and fast construction, etc. Also, AM have benefits in the construction industry under the situation of harsh environment like in an extremely cold environment and high depth mines, which may cause some health issues. In such conditions, off-site construction is preferable in which building components are fabricated by big robotic arms for additive manufacturing and transfers to the actual site. AM also assist in reducing the length of supply chain management by reducing the number of steps [110]. With a large number of benefits, AM offer the freedom in new designs, architects with complexity, high productivity, topological optimization, and waste reduction.

1.5 CONCLUSIONS AND FUTURE SCOPE

Geometrical complexity and customized fabrication have limited the use of conventional manufacturing systems in the modern industries, which is also associated to the lack of sustainability. Due to this, the technique initially developed as the prototyping techniques are now attracting the attention of the engineers to be used in the real manufacturing world. Additive manufacturing is one of such techniques, which is now extensively useful for the fabrication of the customized and complex geometry because of their characteristics of design freedom, volumetric customization, minimal wastes, and functional component fabrication. Hence, a comprehensive review of the several additive manufacturing techniques or 3D-printing methods has been carried out with their recent developments and application in the various areas. The challenges associated with different techniques have also been discussed.

In terms of classification of additive manufacturing techniques, it is done on the basis of states of the input materials, such as liquid, semi-solid, and solids. On these basis, various kinds of AM techniques of different principles like polymerization process, extrusion of fused material, particle sintering, particle melting via laser beam and electron beam, material-jetting, arc welding technique, etc. have been described in this chapter. Among these, the FDM technique is a material extrusion technique of layered fabrication, which is coupled with simplicity and high-rate fabrication with the least post processing. However, the quality of the parts produced is lower than SLS and SLA. SLA is a technique of vat polymerization with very high resolution, which can fabricate the component with a good surface quality and dimensional stability. In the powder bed fusion category, SLS and SLM techniques were considered, which have the resolution a little lower than SLA, but higher than FDM. From the literature, it is confirmed that AM techniques like FDM, SLS, SLA, DIW, 3DP, MJM, DLP, and SC3P were developed initially with the polymeric materials. Subsequently, new AM techniques were developed with the usage of metallic material by the development of SLM, LENS, DED, WAAM, and EBM. In the metallic 3D-printing technologies, different sources of the energy were used like laser, electron beam, electric arc, etc. Furthermore, AM techniques have also been developed for the fabrication of the ceramic materials with various combinations consisting of polymer-ceramic, metal-ceramic, and ceramic-ceramic materials. From the various research articles, it has also been found that functionally graded material components with customized properties have been fabricated by the utilization of additive manufacturing principle for several applications, like aerospace, automotive, biomedical, construction, and energy fields.

In this chapter, a comprehensive study has also been done on the recently developed technique of high resolution like TPP, projection microstereolithography, and electrohydrodynamic printing techniques. The study revealed that the development of such high-resolution technique has provided an opportunity to develop the part from nanoscale to macroscale with high precision. With the development of such techniques, a pathway for AM has been opened in the direction of industrial manufacturing, which can overcome the various limitations of the conventional fabrication techniques. Among these techniques, various applications of different AM techniques have been reported, which are clearly describing the involvement of AM in the industrial revolutions. However, to gain the acceptance in the future of the industries, research and development is essential to elevate the AM over conventional processes.

REFERENCES

[1] B. Berman, "3-D printing: The new industrial revolution," *Bus. Horiz.*, vol. 55, no. 2, pp. 155–162, Mar. 2012, doi: 10.1016/j.bushor.2011.11.003.

[2] J. W. Stansbury and M. J. Idacavage, "3D printing with polymers: Challenges among expanding options and opportunities," *Dent. Mater.*, vol. 32, no. 1, pp. 54–64, Jan. 2016, doi: 10.1016/j.dental.2015.09.018.

[3] P. Wu, J. Wang, and X. Wang, "A critical review of the use of 3-D printing in the construction industry," *Autom. Constr.*, vol. 68, pp. 21–31, Aug. 2016, doi: 10.1016/j.autcon.2016.04.005.

[4] M. Gupta, "3D printing of metals," *Metals (Basel).*, vol. 7, no. 10, pp. 3–4, 2017, doi: 10.3390/met7100403.

[5] O. Ivanova, C. Williams, and T. Campbell, "Additive manufacturing (AM) and nanotechnology: promises and challenges," no. April 2019, 2013, doi: 10.1108/RPJ-12-2011-0127.

[6] S. I. Roohani-Esfahani, P. Newman, and H. Zreiqat, "Design and fabrication of 3D printed scaffolds with a mechanical strength comparable to cortical bone to repair large bone defects," *Sci. Rep.*, vol. 6, Jan. 2016, doi: 10.1038/SREP19468.

[7] M. Vaezi, H. Seitz, and S. Yang, "A review on 3D micro-additive manufacturing technologies," *Int. J. Adv. Manuf. Technol.*, vol. 67, no. 5–8, pp. 1721–1754, 2013, doi: 10.1007/S00170-012-4605-2.

[8] T. D. Ngo, A. Kashani, G. Imbalzano, K. T. Q. Nguyen, and D. Hui, "Additive manufacturing (3D printing): A review of materials, methods, applications and challenges," *Compos. Part B Eng.*, vol. 143, pp. 172–196, Jun. 2018, doi: 10.1016/J.COMPOSITESB.2018.02.012.

[9] B. Bhushan and M. Caspers, "An overview of additive manufacturing (3D printing) for microfabrication," *Microsyst. Technol.*, vol. 23, no. 4, pp. 1117–1124, Apr. 2017, doi: 10.1007/S00542-017-3342-8.

[10] M. Mao *et al.*, "The emerging frontiers and applications of high-resolution 3D printing," *Micromachines*, vol. 8, no. 4, Apr. 2017, doi: 10.3390/MI8040113.

[11] C. Ru, J. Luo, S. Xie, and Y. Sun, "A review of non-contact micro–and nano-printing technologies," *J. Micromechanics Microengineering*, vol. 24, no. 5, p. 053001, Apr. 2014, doi: 10.1088/0960-1317/24/5/053001.

[12] O. A. Mohameds, S. H. Masood, and J. L. Bhowmik, "Optimization of fused deposition modeling process parameters: A review of current research and future prospects," *Adv. Manuf.*, vol. 3, no. 1, pp. 42–53, Mar. 2015, doi: 10.1007/S40436-014-0097-7.

[13] A. K. Sood, R. K. Ohdar, and S. S. Mahapatra, "Parametric appraisal of mechanical property of fused deposition modelling processed parts," *Mater. Des.*, vol. 31, no. 1, pp. 287–295, Jan. 2010, doi: 10.1016/j.matdes.2009.06.016.

[14] J. S. Chohan, R. Singh, K. S. Boparai, R. Penna, and F. Fraternali, "Dimensional accuracy analysis of coupled fused deposition modeling and vapour smoothing operations for biomedical applications," *Compos. Part B Eng.*, vol. 117, pp. 138–149, May 2017, doi: 10.1016/j.compositesb.2017.02.045.

[15] P. Parandoush and D. Lin, "A review on additive manufacturing of polymer-fiber composites," *Compos. Struct.*, vol. 182, pp. 36–53, Dec. 2017, doi: 10.1016/j.compstruct.2017.08.088.

[16] X. Wang, M. Jiang, Z. Zhou, J. Gou, and D. Hui, "3D printing of polymer matrix composites: A review and prospective," *Compos. Part B Eng.*, vol. 110, pp. 442–458, Feb. 2017, doi: 10.1016/j.compositesb.2016.11.034.

[17] T. Sathies, P. Senthil, and M. S. Anoop, "A review on advancements in applications of fused deposition modelling process," *Rapid Prototyp. J.*, vol. 26, no. 4, pp. 669–687, May 2020, doi: 10.1108/RPJ-08-2018-0199/FULL/PDF.

[18] T. Huang, M. S. Mason, G. E. Hilmas, and M. C. Leu, "Freeze-form extrusion fabrication of ceramic parts," doi: 10.1080/17452750600649609, vol. 1, no. 2, pp. 93–100, 2007, doi: 10.1080/17452750600649609.

[19] M. S. Mason, T. Huang, R. G. Landers, M. C. Leu, and G. E. Hilmas, "Aqueous-based extrusion of high solids loading ceramic pastes: Process modeling and control," *J. Mater. Process. Technol.*, vol. 209, no. 6, pp. 2946–2957, Mar. 2009, doi: 10.1016/J.JMATPROTEC.2008.07.004.

[20] M. C. Leu, B. K. Deuser, L. Tang, R. G. Landers, G. E. Hilmas, and J. L. Watts, "Freeze-form extrusion fabrication of functionally graded materials," *CIRP Ann.*, vol. 61, no. 1, pp. 223–226, Jan. 2012, doi: 10.1016/J.CIRP.2012.03.050.

[21] G. Suresh, K. L. Narayana, and M. K. Mallik, "A review on development of medical implants by rapid prototyping technology," *Int. J. Pure Appl. Math.*, vol. 117, no. 21, pp. 257–276, 2017.

[22] I. Gibson, D. Rosen, and B. Stucker, "Sheet lamination processes," *Addit. Manuf. Technol.*, pp. 219–244, 2015, doi: 10.1007/978-1-4939-2113-3_9.

[23] J. Li, T. Monaghan, T. T. Nguyen, R. W. Kay, R. J. Friel, and R. A. Harris, "Multifunctional metal matrix composites with embedded printed electrical materials fabricated by ultrasonic additive manufacturing," *Compos. Part B Eng.*, vol. 113, pp. 342–354, Mar. 2017, doi: 10.1016/j.compositesb.2017.01.013.

[24] B. Mueller and D. Kochan, "Laminated object manufacturing for rapid tooling and patternmaking in foundry industry," *Comput. Ind.*, vol. 39, no. 1, pp. 47–53, 1999, doi: 10.1016/s0166-3615(98)00127-4.

[25] T. Jollivet, A. Darfeuille, B. Verquin, and S. Pillot, "Rapid manufacturing of polymer parts by selective laser sintering," *Int. J. Mater. Form.* 2009 *21*, vol. 2, no. 1, pp. 697–700, Dec. 2009, doi: 10.1007/S12289-009-0604-8.

[26] Y. A. Gueche *et al.*, "Selective laser sintering of solid oral dosage forms with copovidone and paracetamol using a CO_2 laser," *Pharm.* 2021, vol. 13, no. 2, p. 160, Jan. 2021, doi: 10.3390/PHARMACEUTICS13020160.

[27] Y. Bozkurt and E. Karayel, "3D printing technology; methods, biomedical applications, future opportunities and trends," *J. Mater. Res. Technol.*, vol. 14, pp. 1430–1450, Sep. 2021, doi: 10.1016/J.JMRT.2021.07.050.

[28] B. Utela, D. Storti, R. Anderson, and M. Ganter, "A review of process development steps for new material systems in three dimensional printing (3DP)," *Journal of Manufacturing Processes*, vol. 10, no. 2. Elsevier BV, pp. 96–104, Jul. 01, 2008, doi: 10.1016/j.jmapro.2009.03.002.

[29] S. K. Tiwari, S. Pande, S. Agrawal, and S. M. Bobade, "Selection of selective laser sintering materials for different applications," *Rapid Prototyp. J.*, vol. 21, no. 6, pp. 630–648, Oct. 2015, doi: 10.1108/RPJ-03-2013-0027/FULL/PDF.

[30] S. C. Ligon, R. Liska, J. Stampfl, M. Gurr, and R. Mülhaupt, "Polymers for 3D printing and customized additive manufacturing," *Chem. Rev.*, vol. 117, no. 15, pp. 10212–10290, Aug. 2017, doi: 10.1021/ACS.CHEMREV.7B00074/ASSET/IMAGES/LARGE/CR-2017-00074G_0037.JPEG.

[31] H. Meier and C. Haberland, "Experimental studies on selective laser melting of metallic parts," *Materwiss. Werksttech.*, vol. 39, no. 9, pp. 665–670, Sep. 2008, doi: 10.1002/MAWE.200800327.

[32] D. Wang, Y. Yang, R. Liu, D. Xiao, and J. Sun, "Study on the designing rules and processability of porous structure based on selective laser melting (SLM)," *J. Mater. Process. Technol.*, vol. 213, no. 10, pp. 1734–1742, 2013, doi: 10.1016/J.JMATPROTEC.2013.05.001.

[33] J. Zhang, B. Song, Q. Wei, D. Bourell, and Y. Shi, "A review of selective laser melting of aluminum alloys: Processing, microstructure, property and developing trends," *J. Mater. Sci. Technol.*, vol. 35, no. 2, pp. 270–284, Feb. 2019, doi: 10.1016/J.JMST.2018.09.004.

[34] C. Y. Yap *et al.*, "Review of selective laser melting: Materials and applications," *Appl. Phys. Rev.*, vol. 2, no. 4, Dec. 2015, doi: 10.1063/1.4935926.

[35] R. Acharya, R. Bansal, J. J. Gambone, and S. Das, "A microstructure evolution model for the processing of single-crystal alloy CMSX-4 through scanning laser epitaxy for turbine engine hot-section component repair (part II)," *Metall. Mater. Trans. B Process Metall. Mater. Process. Sci.*, vol. 45, no. 6, pp. 2279–2290, Dec. 2014, doi: 10.1007/S11663-014-0183-Z/FIGURES/14.

[36] O. Abdulhameed, A. Al-Ahmari, W. Ameen, and S. H. Mian, "Additive manufacturing: Challenges, trends, and applications:," doi: 10.1177/1687814018822880, vol. 11, no. 2, pp. 1–27, Feb. 2019, doi: 10.1177/1687814018822880.

[37] M. Xia, B. Nematollahi, and J. G. Sanjayan, "Development of powder-based 3D boncrete printing using geopolymers," in *3D Concrete Printing Technology*, Butterworth-Heinemann, 2019, pp. 223–240.

[38] B. Nematollahi, M. Xia, and J. Sanjayan, "Post-processing methods to improve strength of particle-bed 3d printed geopolymer for digital construction applications," *Front. Mater.*, vol. 6, p. 160, Jul. 2019, doi: 10.3389/FMATS.2019.00160/BIBTEX.

[39] M. Galati and L. Iuliano, "A literature review of powder-based electron beam melting focusing on numerical simulations," *Addit. Manuf.*, vol. 19, pp. 1–20, Jan. 2018, doi: 10.1016/J.ADDMA.2017.11.001.

[40] M. F. Zäh and • S Lutzmann, "PRODUCTION PROCESS Modelling and simulation of electron beam melting," doi: 10.1007/s11740-009-0197-6.

[41] P. Heinl, A. Rottmair, C. Körner, and R. F. Singer, "Cellular titanium by selective electron beam melting," *Adv. Eng. Mater.*, vol. 9, no. 5, pp. 360–364, May 2007, doi: 10.1002/ADEM.200700025.

[42] C. Körner, "Additive manufacturing of metallic components by selective electron beam melting — a review," vol. 61, no. 5, pp. 361–377, 2016, doi: 10.1080/09506608.2016.1176289.

[43] M. Izadi, A. Farzaneh, M. Mohammed, I. Gibson, and B. Rolfe, "A review of laser engineered net shaping (LENS) build and process parameters of metallic parts," *Rapid Prototyp. J.*, vol. 26, no. 6, pp. 1059–1078, Jun. 2020, doi: 10.1108/RPJ-04-2018-0088/FULL/PDF.

[44] R. R. Unocic and J. N. DuPont, "Process efficiency measurements in the laser engineered net shaping process," *Metall. Mater. Trans. B* 2004 351, vol. 35, no. 1, pp. 143–152, 2004, doi: 10.1007/S11663-004-0104-7.

[45] I. Palčič, M. Balažic, M. Milfelner, and B. Buchmeister, "Potential of Laser Engineered Net Shaping (LENS) Technology," https://doi.org/10.1080/10426910902809776, vol. 24, no. 7–8, pp. 750–753, Jul. 2009, doi: 10.1080/10426910902809776.

[46] R. Rumman, D. A. Lewis, J. Y. Hascoet, and J. S. Quinton, "Laser metal deposition and wire arc additive manufacturing of materials: An overview," *Arch. Met. Mater*, vol. 64, pp. 467–473, 2019, doi: 10.24425/amm.2019.127561.

[47] Q. Liu *et al.*, "Microstructure and mechanical properties of LMD-SLM hybrid forming Ti6Al4V alloy," 2016, doi: 10.1016/j.msea.2016.02.069.

[48] C. Cavallo, "All about laser metal deposition 3D printing." https://www.thomasnet.com/articles/custom-manufacturing-fabricating/all-about-laser-metal-deposition-3d-printing/ (accessed Sep. 05, 2022).

[49] M. Chaturvedi, E. Scutelnicu, C. C. Rusu, L. R. Mistodie, D. Mihailescu, and S. Arungalai Vendan, "Wire arc additive manufacturing: Review on recent findings and challenges in industrial applications and materials characterization," vol. 11, no. 6, p. 939, Jun. 2021, doi: 10.3390/MET11060939.

[50] B. Wu *et al.*, "A review of the wire arc additive manufacturing of metals: properties, defects and quality improvement," *J. Manuf. Process.*, vol. 35, pp. 127–139, Oct. 2018, doi: 10.1016/J.JMAPRO.2018.08.001.

[51] T. Chang, X. Fang, G. Liu, H. Zhang, and K. Huang, "Wire and arc additive manufacturing of dissimilar 2319 and 5B06 aluminum alloys," *J. Mater. Sci. Technol.*, vol. 124, pp. 65–75, Oct. 2022, doi: 10.1016/J.JMST.2022.02.024.

[52] K. Hartmann *et al.*, "Robot-assisted shape deposition manufacturing," *Proc.–IEEE Int. Conf. Robot. Autom.*, no. pt 4, pp. 2890–2895, 1994, doi: 10.1109/ROBOT.1994.350900.

[53] A. Shen, D. Caldwell, A. W. K. Ma, and S. Dardona, "Direct write fabrication of high-density parallel silver interconnects," *Addit. Manuf.*, vol. 22, pp. 343–350, Aug. 2018, doi: 10.1016/J.ADDMA.2018.05.010.

[54] J. Malda *et al.*, "25th Anniversary Article: Engineering Hydrogels for Biofabrication," *Adv. Mater.*, vol. 25, no. 36, pp. 5011–5028, Sep. 2013, doi: 10.1002/ADMA.201302042.

[55] R. D. Farahani, M. Dubé, and D. Therriault, "Three-dimensional printing of multifunctional nanocomposites: Manufacturing techniques and applications," *Adv. Mater.*, vol. 28, no. 28, pp. 5794–5821, Jul. 2016, doi: 10.1002/ADMA.201506215.

[56] C. Zhu *et al.*, "Supercapacitors based on three-dimensional hierarchical graphene aerogels with periodic macropores," *Nano Lett.*, vol. 16, no. 6, pp. 3448–3456, Jun. 2016, doi: 10.1021/ACS.NANOLETT.5B04965/SUPPL_FILE/NL5B04965_SI_003.MOV.

[57] B. Wang, Z. Zhang, Z. Pei, J. Qiu, and S. Wang, "Current progress on the 3D printing of thermosets," *Adv. Compos. Hybrid Mater.*, vol. 3, no. 4, pp. 462–472, Dec. 2020, doi: 10.1007/S42114-020-00183-Z/FIGURES/9.

[58] C. Xu, A. Bouchemit, E. Rance, L. Laberge Lebel, and D. Therriault, "Solvent-cast based metal 3D printing and secondary metallic infiltration," *J. Mater. Chem. C*, vol. 5, p. 10448, doi: 10.1039/c7tc02884a.

[59] J. Singh, P. M. Pandey, T. Kaur, and N. Singh, "Effect of heparin drug loading on biodegradable polycaprolactone–iron pentacarbonyl powder blend stents fabricated by solvent cast 3D printing," *Rapid Prototyp. J.*, 2022, doi: 10.1108/RPJ-02-2021-0043/FULL/PDF.

[60] J. Singh, P. M. Pandey, T. Kaur, and N. Singh, "A comparative analysis of solvent cast 3D printed carbonyl iron powder reinforced polycaprolactone polymeric stents for intravascular applications," *J. Biomed. Mater. Res. Part B Appl. Biomater.*, pp. 1–16, 2021, doi: 10.1002/jbm.b.34795.

[61] J. Singh, G. Singh, and P. M. Pandey, "Multi-objective optimization of solvent cast 3D printing process parameters for fabrication of biodegradable composite stents," *Int. J. Adv. Manuf. Technol.*, vol. 115, no. 11–12, pp. 3945–3964, 2021, doi: 10.1007/s00170-021-07423-6.

[62] S. Z. Guo, F. Gosselin, N. Guerin, A. M. Lanouette, M. C. Heuzey, and D. Therriault, "Solvent-cast three-dimensional printing of multifunctional microsystems," *Small*, vol. 9, no. 24, pp. 4118–4122, 2013, doi: 10.1002/smll.201300975.

[63] J. Singh, P. M. Pandey, T. Kaur, and N. Singh, "Surface characterization of polycaprolactone and carbonyl iron powder composite fabricated by solvent cast 3D printing for tissue engineering," *Polym. Compos.*, pp. 1–7, 2020, doi: 10.1002/pc.25871.

[64] J. Singh, T. Kaur, N. Singh, and P. M. Pandey, "Biological and mechanical characterization of biodegradable carbonyl iron powder/polycaprolactone composite material fabricated using three-dimensional printing for cardiovascular stent application," *Proc. Inst. Mech. Eng. Part H J. Eng. Med.*, vol. 234, no. 9, pp. 975–987, 2020, doi: 10.1177/0954411920936055.

[65] F. P. W. Melchels, J. Feijen, and D. W. Grijpma, "A review on stereolithography and its applications in biomedical engineering," *Biomaterials*, vol. 31, no. 24, pp. 6121–6130, Aug. 2010, doi: 10.1016/j.biomaterials.2010.04.050.

[66] A. Razavykia, E. Brusa, C. Delprete, and R. Yavari, "An overview of additive manufacturing technologies—a review to technical synthesis in numerical study of selective laser melting," *Mater. 2020, Vol. 13, Page 3895*, vol. 13, no. 17, p. 3895, Sep. 2020, doi: 10.3390/MA13173895.

[67] H. Lee, C. H. J. Lim, M. J. Low, N. Tham, V. M. Murukeshan, and Y. J. Kim, "Lasers in additive manufacturing: A review," *Int. J. Precis. Eng. Manuf.–Green Technol.*, vol. 4, no. 3, pp. 307–322, Jul. 2017, doi: 10.1007/S40684-017-0037-7.

[68] Y. Lu, G. Mapili, G. Suhali, S. Chen, and K. Roy, "A digital micro-mirror device-based system for the microfabrication of complex, spatially patterned tissue engineering scaffolds," *J. Biomed. Mater. Res. A*, vol. 77, no. 2, pp. 396–405, May 2006, doi: 10.1002/JBM.A.30601.

[69] J. Zhang, Q. Hu, S. Wang, J. Tao, and M. Gou, "Digital light processing based three-dimensional printing for medical applications," *Int. J. Bioprinting*, vol. 6, no. 1, pp. 12–27, 2020, doi: 10.18063/ijb.v6i1.242.

[70] K. Kitsakis, J. Kechagias, N. Vaxevanidis, and D. Giagkopoulos, "Tolerance Analysis of 3d-MJM parts according to IT grade," in *IOP Conference Series: Material Science and Engineering (161)*, 2016, pp. 1–11, doi: 10.1088/1757-899X/161/1/012024.

[71] "Additive manufacturing: Challenges, trends, and applications–Osama Abdulhameed, Abdulrahman Al-Ahmari, Wadea Ameen, Syed Hammad Mian, 2019." https://journals.sagepub.com/doi/full/10.1177/1687814018822880 (accessed Jun. 23, 2022).

[72] "Rapid Prototyping: Principles And Applications (3rd Edition) (With Companio...–Google Books." https://www.google.co.in/books/edition/Rapid_Prototyping_Principles_And_Applica/PiI8DQAAQBAJ?hl=en&gbpv=0 (accessed Jun. 02, 2022).

[73] P. P. Kamble, S. Chavan, R. Hodgir, G. Gote, and K. P. Karunakaran, "Multi-jet ice 3D printing," *Rapid Prototyp. J.*, vol. 28, no. 6, pp. 989–1004, May 2022, doi: 10.1108/RPJ-03-2021-0065/FULL/PDF.

[74] G. B. Kim *et al.*, "Three-dimensional printing: Basic principles and applications in medicine and radiology," *Korean J. Radiol.*, vol. 17, no. 2, pp. 182–197, Mar. 2016, doi: 10.3348/KJR.2016.17.2.182.

[75] H. H. Hwang, W. Zhu, G. Victorine, N. Lawrence, and S. Chen, "3D-printing of functional biomedical microdevices via light- and extrusion-based approaches," *Small Methods*, vol. 2, no. 2, p. 1700277, Feb. 2018, doi: 10.1002/SMTD.201700277.

[76] W. Zhang, M. C. Leu, Z. Ji, and Y. Yan, "Rapid freezing prototyping with water," *Mater. Des.*, vol. 20, no. 2–3, pp. 139–145, Jun. 1999, doi: 10.1016/S0261-3069(99)00020-5.

[77] W. Zhang, M. C. Leu, Z. Ji, and Y. Yan, "US6253116B1–Method and apparatus for rapid freezing prototyping–Google Patents," 2001.

[78] "The freezing-point, boiling-point and conductivity methods," *Nat. 1898 571487*, vol. 57, no. 1487, pp. 606–606, Apr. 1898, doi: 10.1038/057606b0.

[79] T. Wohlers, "Developments in additive manufacturing," *Manuf. Eng.*, vol. 144, no. 1, pp. 37–62, Jan. 2021, doi: 10.1016/B978-0-12-822056-6.00002-3.

[80] Z. Liu *et al.*, "Study on chemical graft structure modification and mechanical properties of photocured polyimide," *ACS Omega*, vol. 7, no. 11, pp. 9582–9593, Mar. 2022, doi: 10.1021/ACSOMEGA.1C06933.

[81] Y. Wang *et al.*, "Two-photon excited deep-red and near-infrared emissive organic co-crystals," *Nat. Commun. 2020 111*, vol. 11, no. 1, pp. 1–11, Sep. 2020, doi: 10.1038/s41467-020-18431-7.

[82] A. J. G. Otuka, N. B. Tomazio, K. T. Paula, and C. R. Mendonça, "Two-photon polymerization: functionalized microstructures, micro-resonators, and bio-scaffolds," *Polym. 2021, Vol. 13, Page 1994*, vol. 13, no. 12, p. 1994, Jun. 2021, doi: 10.3390/POLYM13121994.
[83] V. Harinarayana and Y. C. Shin, "Two-photon lithography for three-dimensional fabrication in micro/ nanoscale regime: A comprehensive review," *Opt. Laser Technol.*, vol. 142, 2021, doi: 10.1016/j.optlastec.2021.107180.
[84] A. Z. Zabidi *et al.*, "Computational mechanical characterization of geometrically transformed Schwarz P lattice tissue scaffolds fabricated via two photon polymerization (2PP)," *Addit. Manuf.*, vol. 25, pp. 399–411, Jan. 2019, doi: 10.1016/J.ADDMA.2018.11.021.
[85] J. U. Park *et al.*, "High-resolution electrohydrodynamic jet printing," *Nat. Mater. 2007 610*, vol. 6, no. 10, pp. 782–789, Aug. 2007, doi: 10.1038/nmat1974.
[86] M. S. Onses, E. Sutanto, P. M. Ferreira, A. G. Alleyne, and J. A. Rogers, "Mechanisms, capabilities, and applications of high-resolution electrohydrodynamic jet printing," *Small*, vol. 11, no. 34, pp. 4237–4266, Sep. 2015, doi: 10.1002/SMLL.201500593.
[87] Y. Han and J. Dong, "Electrohydrodynamic printing for advanced micro/ nanomanufacturing: Current progresses, opportunities, and challenges," *J. Micro Nano-Manufacturing*, vol. 6, no. 4, pp. 1–20, 2018, doi: 10.1115/1.4041934.
[88] J. B. Fenn, M. Mann, C. K. Meng, S. F. Wong, and C. M. Whitehouse, "Electrospray ionization–principles and practice," *Mass Spectrom. Rev.*, vol. 9, no. 1, pp. 37–70, Jan. 1990, doi: 10.1002/MAS.1280090103.
[89] A. Jaworek and A. Krupa, "Jet and drops formation in electrohydrodynamic spraying of liquids. A systematic approach," *Exp. Fluids*, vol. 27, no. 1, pp. 43–52, 1999, doi: 10.1007/S003480050327.
[90] Q. Lei, J. He, and D. Li, "Electrohydrodynamic 3D printing of layer-specifically oriented, multiscale conductive scaffolds for cardiac tissue engineering," *Nanoscale*, vol. 11, no. 32, pp. 15195–15205, Aug. 2019, doi: 10.1039/C9NR04989D.
[91] C. Sun, N. Fang, D. M. Wu, and X. Zhang, "Projection micro-stereolithography using digital micro-mirror dynamic mask," *Sensors Actuators, A Phys.*, vol. 121, no. 1, pp. 113–120, May 2005, doi: 10.1016/J.SNA.2004.12.011.
[92] A.Bertsch,S.Zissi,J.Y.Jézéquel,S.Corbel,andJ.C.André,"Microstereophotolithography using a liquid crystal display as dynamic mask-generator," *Microsyst. Technol.*, vol. 3, no. 2, pp. 42–47, 1997, doi: 10.1007/S005420050053.
[93] R. Raman *et al.*, "High-resolution projection microstereolithography for patterning of neovasculature," *Adv. Healthc. Mater.*, vol. 5, no. 5, pp. 610–619, Mar. 2016, doi: 10.1002/ADHM.201500721.
[94] B. Ahuja, M. Karg, and M. Schmidt, "Additive manufacturing in production: challenges and opportunities," vol. 9353, pp. 11–20, Mar. 2015, doi: 10.1117/12.2082521.
[95] "Norsk Titanium | Norsk Titanium to Deliver the World's First FAA-Approved, 3D-Printed, Structural Titanium Components to Boeing." https://www.norsktitanium.com/media/press/norsk-titanium-to-deliver-the-worlds-first-faa-approved-3d-printed-structural-titanium-components-to-boeing (accessed Jul. 04, 2022).
[96] "GKN Aerospace delivers revolutionary Ariane 6 Nozzle to Airbus Safran Launchers." https://www.gknaerospace.com/en/newsroom/news-releases/2017/gkn-delivers-revolutionary-ariane-6-nozzle-to-airbus-safran-launchers/ (accessed Jul. 04, 2022).
[97] "Fit to Print: New Plant Will Assemble World's First Passenger Jet Engine With 3D Printed Fuel Nozzles, Next-Gen Materials | GE News." https://www.ge.com/news/reports/fit-to-print (accessed Jul. 04, 2022).

[98] "A Fully Non-Metallic Gas Turbine Engine Enabled by Additive Manufacturing Part I: System Analysis, Component Identification, Additive Manufacturing, and Testing of Polymer Composites–NASA Technical Reports Server (NTRS)." https://ntrs.nasa.gov/citations/20150010717 (accessed Jun. 02, 2022).

[99] A. Zocca, P. Colombo, C. M. Gomes, and J. Günster, "Additive manufacturing of ceramics: issues, potentialities, and opportunities," *J. Am. Ceram. Soc.*, vol. 98, no. 7, pp. 1983–2001, Jul. 2015, doi: 10.1111/JACE.13700.

[100] R. Chang, K. Emami, H. Wu, and W. Sun, "Biofabrication of a three-dimensional liver micro-organ as an in vitro drug metabolism model," *Biofabrication*, vol. 2, no. 4, p. 045004, Nov. 2010, doi: 10.1088/1758-5082/2/4/045004.

[101] J. Giannatsis and V. Dedoussis, "Dedoussis. Additive fabrication technologies applied to medicine and health care: A review," *Int. J. Adv. Manuf. Technol.*, vol. 40, no. 1–2, pp. 116–127, Jan. 2009, doi: 10.1007/s00170-007-1308-1.

[102] J. Singh and P. M. Pandey, "Solvent cast 3D printing of bioresorbable composite stents: analysis of process parameters for geometry dimensions and mechanical properties optimization," *Addit. Manuf.*, p. (Under review).

[103] J. Singh, P. M. Pandey, T. Kaur, and N. Singh, "Effect of heparin drug loading on biodegradable polycaprolactone–iron pentacarbonyl powder blend stents fabricated by solvent cast 3D printing," *Biofabrication*, p. (Under Review).

[104] J. Singh, T. Kaur, N. Singh, and P. M. Pandey, "Biological and mechanical characterization of biodegradable carbonyl iron powder/polycaprolactone composite material fabricated using three-dimensional printing for cardiovascular stent application," *Proc. Inst. Mech. Eng. Part H J. Eng. Med.*, vol. 234, no. 9, pp. 975–987, 2020.

[105] J. Singh, H. Singh, and U. Batra, "Magnesium doped hydroxyapatite: synthesis, characterization and bioactivity evaluation," in *Biomaterials Science: Processing, Properties and Applications V: Ceramic Transactions, Volume 254*, 2015.

[106] Q. Liu, M. C. Leu, and S. M. Schmitt, "Rapid prototyping in dentistry: technology and application," *Int. J. Adv. Manuf. Technol.* vol. 29, no. 3, pp. 317–335, Aug. 2005, doi: 10.1007/S00170-005-2523-2.

[107] L. Dall'Ava, H. Hothi, A. Di Laura, J. Henckel, and A. Hart, "3D printed acetabular cups for total hip arthroplasty: A review article," *Met.* 2019, vol. 9, no. 7, p. 729, Jun. 2019, doi: 10.3390/MET9070729.

[108] Y. Dong, C. Bao, and W. S. Kim, "Sustainable additive manufacturing of printed circuit Boards," *Joule*, vol. 2, no. 4, pp. 579–582, Apr. 2018, doi: 10.1016/J.JOULE.2018.03.015.

[109] Z. Liu, M. Zhang, B. Bhandari, and Y. Wang, "3D printing: Printing precision and application in food sector," *Trends Food Sci. Technol.*, vol. 69, pp. 83–94, Nov. 2017, doi: 10.1016/J.TIFS.2017.08.018.

[110] Y. Huang, M. C. Leu, J. Mazumder, and A. Donmez, "Additive manufacturing: Current state, future potential, gaps and needs, and recommendations," *J. Manuf. Sci. Eng. Trans. ASME*, vol. 137, no. 1, Feb. 2015, doi: 10.1115/1.4028725/375256.

2 An Overview of Post-Processing Technologies in General Practice

Haitao Zhu, Allan E.W. Rennie, and Yingtao Tian
School of Engineering, Lancaster University, Bailrigg, Lancaster, UK

CONTENTS

2.1 INTRODUCTION

Additive Manufacturing (AM) forms components layer-by-layer, providing the advantages of generating less material waste, support of complex structures, etc. However, to satisfy the functional, geometrical and/or dimensional requirements of final products, a series of subtractive machining, surface modification, thermal and non-thermal treatments may need to be introduced. The processing procedure after component removal from the working platform is known as post-processing, which plays a fundamental role and accounts for between 4-13% of the cost of the whole AM process (Thomas and Gilbert 2014).

Currently, most machining technologies, such as mechanical machining, electrochemical machining, laser machining, etc., and other processing technologies have been proven to be effective in improving the surface and bulk properties of the AM components. However, the need to appropriately select and design ideal post-processing technologies and procedures should comply with the AM process and product requirements. For example, chemical and electrochemical machining could

DOI: 10.1201/9781003276111-2

be ideal methods for polishing complex-shaped components. Mechanical machining could be used as pre-treatment for surface finishing before the electrochemical polishing when requiring a high roughness reduction (e.g., from 20 μm to below 1 μm). Thermal treatment with different temperatures and cooling methods could provide components with different ductility, fatigue life, hardness, etc. In this case, practitioners who can offer products satisfying the customers' expectations and cost savings will be more competitive. This requires practitioners to be knowledgeable not only about AM processes but also have a working appreciation of post-processing technologies that could be deployed.

This chapter aims to provide an overview of common post-processing technologies from the perspective of utilization so that readers have a basic cognition of what and why components should be post-processed, what the general post-processing procedure is, and the common post-processing technologies in general practice. The chapter is divided into four sections, including surface cleaning, surface modification, dimensional accuracy, and mechanical property improvement. The first section introduces the removal of excess powder and support structures (where they are required). The second section introduces surface polishing, protection, functionalization, and a summary of corresponding technologies. Since machining is a materials removal process and some surface protection/functionalization is a materials addition process, the dimensional accuracy and geometric structure could be influenced during the processing. Therefore, the third section introduces the factors that should be fully considered during component fabrication and post-processing. The final section introduces some non-thermal and thermal technologies to improve the mechanical properties of components.

2.2 EXCESS POWDER AND SUPPORT STRUCTURE REMOVAL

AM technologies that use powder materials as the feedstock, should ensure that residual powders are removed from surfaces, holes, cavities, and other geometric features. In the vast majority of situations, most of this excess powder can be removed using brushes, compressed airflow, and/or sandblasting, or for significant residues, removal using advanced cleaning methods, such as pressure washing, ultrasonic cleaning, plasma cleaning, chemical cleaning, etc.

As a consequence of the way that AM technologies join materials together layer by layer, support structures may have to be designed such that the overlying layers vary from the footprint. This is a fundamental issue associated with AM technologies and model complexity. For example, no support structures are required for the selective laser sintering (SLS) process because the powder bed could provide support functions in the manufacturing process, whilst direct metal laser sintering (DMLS) still requires support structures to anchor parts to the build platen and avoid thermal deformation, although the powder bed is used in the same way. Stereolithography (SLA) always requires support structures to ensure the fix of the parts to the build platform. Normally, any overhangs of less than 45° can be manufactured without support structures in DMLS and fusion deposition modeling (FDM), whilst all overhangs require support structures, regardless of the angle for material jetting (MJ)

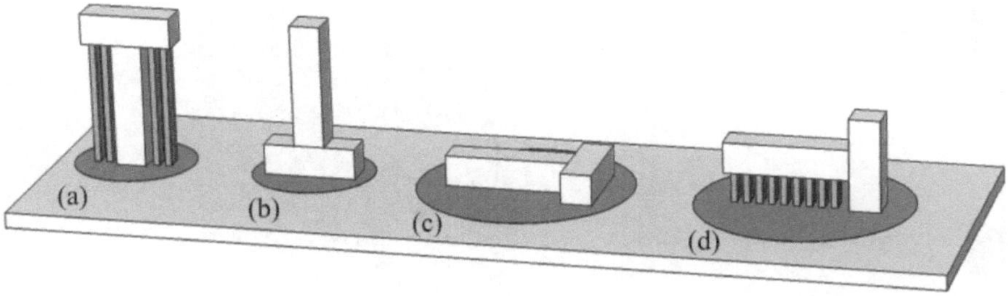

FIGURE 2.1 Four orientations for printing the word 'T'.

technologies. The demand and design of the support structure will influence the final quality and the subsequent cost of manufacture of the final product (Jiang, Xu, and Stringer 2018).

One factor that determines manufacturing cost, quality, and the difficulty of the removal process is the form of the materials used for the support structure (Thomas and Gilbert 2014). As with the fabricated component, these support structures may be of the same or a different material. Polymer materials have a low hardness that allows for removal manually or with simple tools, whereas metals and ceramics would require removing with more complex tools, complemented by sandblasting, wire cutting etc. Additionally, the support structure material often has features that are readily soluble in certain solutions, such as water-based or chemical solutions, or they have a low melting point or shore hardness. Noticeably, the solutions must not influence the base AM components, and as best possible, the removal process should try to eliminate leaving witness marks and rough areas on the component surfaces.

Other determining factors include the position and orientation of the parts within the build envelope of the AM technology and, where appropriate, the influence this has on generating support structures (Unkovskiy et al. 2018). Figure 2.1 presents four orientations for fabricating the letter 'T', where Figure 2.1(a) and (d) require support structure and thus will consume more material in the manufacturing process. Figure 2.1(c) requires more materials for manufacturing the raft because there is more contact with the platform than with Figure 2.1(b). Noticeably, the raft geometry might not be encapsulating and should follow the geometry of the parts in the real manufacturing process.

2.3 SURFACE MODIFICATION

2.3.1 Surface Finish

Although surface conditions of manufactured parts vary depending on the AM technique used, in most cases, they typically have rough surfaces with micro-scale burrs, stair-step structures, spherical protruding features or small cavities (Lou et al. 2019; Sanaei and Fatemi 2021). Figure 2.2 displays the surface topography of a selective laser melting (SLM) 316L stainless steel sample manufactured by

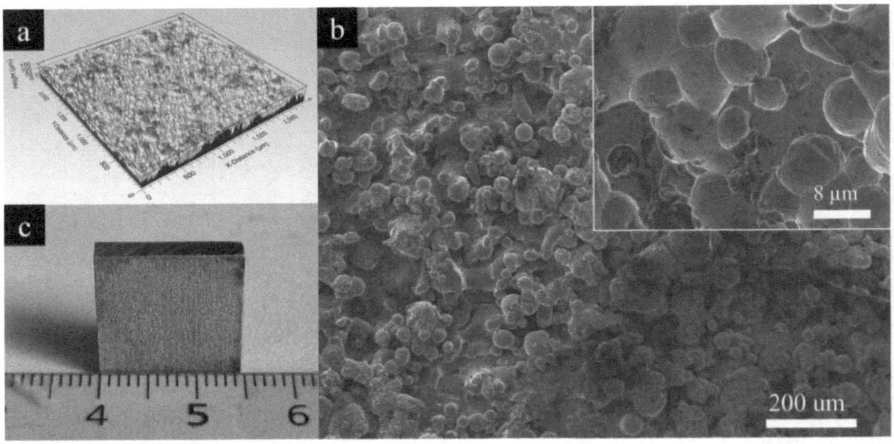

FIGURE 2.2 Surface topography of an SLM 316L stainless steel after removing from platform (a: optical microscope, b: SEM, c: physical sample) (Zhu et al. 2022).

TABLE 2.1
Surface Roughness of Parts Manufactured by Different AM Technologies

AM Techniques	Roughness (Sa)
Vat photopolymerization	> 0.5 μm (Ra)
Selective laser melting	> 4 μm
Electron-beam melting	> 25 μm
Binder jetting	> 10 μm
Laminated object manufacturing	> 10 μm
Directed energy deposition	> 10 μm

an EOS M209 machine with a laser power of 220 W, scanning speed of 700 mm/s, layer thickness of 0.04 mm, and hatch distance of 0.11 mm. The sample was cut into 15 × 15 mm blocks and the as-built surface roughness was measured to be approximately 14 μm using a LEXT OLY5000 microscope and FESEM (Zhu et al. 2022). Many residuals and unsintered powder particles are embedded in the surfaces and cover the melted tracks. General surface roughness for parts manufactured with AM techniques is summarized in Table 2.1 (Santos et al. 2022; Ni et al. 2020; Shanbhag et al. 2021; Zhang et al. 2021; Smith et al. 2021; Ribeiro, Mariani, and Coelho 2020).

Polymer components usually have a low hardness that allows them to be polished efficiently with an abrasive media (e.g., sandpaper). However, for complex-shaped parts, mechanical polishing is limited due to the tool's shapes and sizes. Abrasive flow processes can be used to polish the interior surfaces of workpieces, but are similarly limited by the geometric complexity of the parts and their size (Dixit, Sharma, and Kumar 2021). Metals and ceramics can also be improved with sandpaper or abrasive

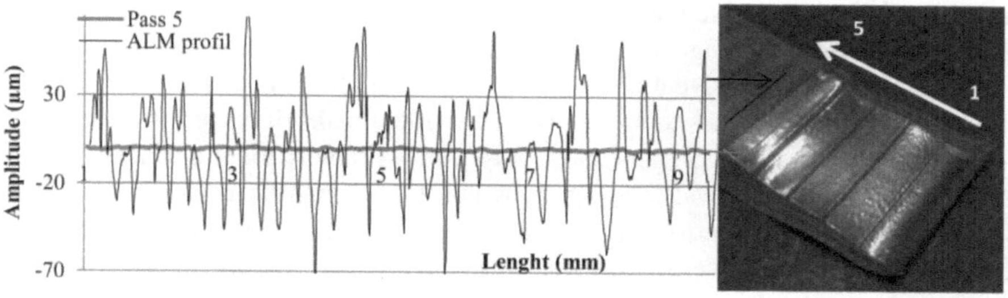

FIGURE 2.3 Profile roughness before and after laser polishing (1–5 represent the passes label and the time that the laser polishing is applied on the surface) (Rosa, Mognol, and Hascoët 2015).

methods; however, the process may require a significant investment of time to obtain the desired surface effects and the dimensional accuracy may be low. Figure 2.3 shows the surface profile before and after the laser polishing of 316L stainless steel components. The amplitude difference of the surface profile almost disappears and the surface becomes smooth and more reflective with additional time spent laser polishing on the surface (Rosa, Mognol, and Hascoët 2015).

2.3.2 Functionalisation and Surface Protection

Products applied in aerospace, automotive, marine, or other extreme environments demand high standards of corrosion and wear resistance. Surface machining approaches make no significant improvement to these properties, and surface coating or alloying techniques should therefore be employed (Maleki et al. 2021). For example, coating stainless steel with a layer of chromium improves surface hardness and corrosion resistance from alkalis, sulphides, nitric acid, and most organic acids. As titanium and its alloys are widely employed in the production of medical implants, plasma surface treatment processes for sterilizing and improving the surface wettability is crucial before applying the implant to the human body.

2.3.3 Strategy

Surface post-treatment strategies should be determined by material, geometry, manufacturing techniques, and customer expectations. Nowadays, an increasing number of surface treatment techniques have been proven suitable for the post-processing of AM components. Table 2.2 summarizes some common post-processing strategies (Gibson et al. 2021; Gupta and Pramanik 2021).

Conventional machining techniques are suitable for processing most materials and producing pieces with high surface quality and dimensional accuracy. However, they cannot process complex-shaped parts because the size and geometry of tools limit their access to interior surfaces. Thermal-based approaches apply high-power optic or electron beams to remelt the workpiece surface, which enables the machining of AM parts with extreme hardness or brittle materials, however, they cannot access the internal surface, which is blocked by the exterior surfaces. Abrasive-based approaches enable finishing by removing materials through friction between the AM components and the abrasive material, resulting in micron and nano surface finishes, smoothing internal surfaces to an almost mirror-like finish. However, this technique cannot polish large-sized components or those with extremely rough surfaces because of the low material removal rate, which would take a significant amount of time. Chemical processing removes materials through chemical reactions between the components and electrolytes, which can render all surfaces smooth, regardless of the geometry. However, the polishing time for rough surfaces is lengthy and can depend significantly on the electrolyte composition. For example, polishing titanium or its alloys requires solutions containing hydrofluoric acid that are toxic. Electrochemical and plasma electrolytic methods can reduce the dependence on electrolyte composition and polish metal parts efficiently whereas surfaces that are inaccessible by the cathode cannot be polished. Electrical discharge machining (EDM) works similarly to plasma electrolytic machining in that a large potential is applied between the electrodes (the tool acting as the anode and the component as the cathodic workpiece). The difference is that EDM employs a dielectric solution in which the workpiece or component is submerged, whereas plasma electrolytic processing employs aqueous solutions of salt.

Some of the aforementioned methods can be applied simultaneously to improve processing effects. For example, smooth rotation cathodes can be employed in the electrochemical polishing process to remove long-wave features from the surface of the workspace (ECMP – electrochemical mechanical polishing).

Surface coating is the most common method to differentiate surface characteristics from bulk materials for protection and functionalization. Spraying technologies work with a wide range of coating materials including metals, alloys, ceramic, plastic, and polymer. It has the advantages of high efficiency, low waste, applicable for thick coating, etc. However, for interior surfaces, the spraying uniformity would be worse and sometimes a manual repair is required. Additionally, dust would be a serious issue during the spraying process. Chemical oxidation can improve the surface ability of anti-corrosion by forming a dense oxide film, which suits metals, but the effect highly relies on the metal oxide itself. Electroplating restores and coats the metal ions in the electrolyte on the product surfaces, providing the surface characteristic of coated metals. These technologies can process complex-shaped components and provide surface films with varied thicknesses but takes time. Surface alloying can be an advanced alternative to electroplating technologies to provide surface layers with better uniformity, stronger binding, higher hardness, and lower cost. The layer generated is composed of amorphous alloy by reacting with metal halide vapour, which is different from the electroplating of pure metals. Physical and chemical vapor deposition provide similar coating effects (thin coating layer with high anti-corrosion properties, etc.) although the operation windows are different. Ion deposition oxide is a high-precision technology that can control the thickness of the coating layer, and the results are repeatable. High-temperature oxidation is a simple process to generate metal oxide on the sample surfaces under high temperatures.

2.4 MANUFACTURING ACCURACY

The similarity in size and geometry of the designed and fabricated parts can be referred to as manufacturing dimensional accuracy. Although current AM technologies may not be as accurate as CNC subtractive machining processes, it is possible to make improvements by adjusting part size, manufacturing speed, material feed rate, temperature control, etc. Potential factors that may affect manufacturing accuracy include the following:

- complexity, resolution, and dimensions of the model in the design and component slicing process;
- inherent accuracy of the AM machine during component fabrication;
- manufacturing parameters including fabrication speed, material feed rate, layer thickness, etc;
- material composition, temperature distribution, etc. during the manufacturing process;
- surface machining during post-processing.

In order to mitigate these factors, the following actions can be taken to improve the manufacturing process:

- complex features should be eliminated or simplified where possible, and design in suitable support structures to avoid mechanical or thermal distortion;
- *.stl files should be exported with as high a resolution as is practicably possible;
- AM machines should be calibrated before manufacturing high-precision parts;
- distance between the working plate/platform and the material deposition nozzle (e.g., in the case of FDM processes) should be set appropriately based on the layer thickness applied;
- decrease the manufacturing speed and choose an appropriate feed rate to avoid internal pores or swell;
- choose appropriate materials and compensate for any deformation due to shrinkage and residual stress by applying scaling factors within CAD models;
- adopt a heated working plate/platform or temperature-controlled chamber to reduce the thermal gradient;
- add extra material (machining allowances) and use subtractive post-processing machining methods (as listed in Table 2.2).

2.5 MECHANICAL PROPERTIES IMPROVEMENT

As shown in Figure 2.4, porosity and distortions may be created during the manufacturing process, which can affect the mechanical properties and ultimately the performance of AM components (Tian et al. 2020). In addition, the application environment may have requirements for mechanical properties such as hardness, elasticity, ductility, etc.

TABLE 2.2
Common Machining Strategies

Functions	Types	Examples
Subtractive machining (Gibson et al. 2021; Gupta and Pramanik 2021)	Conventional methods	Grinding Turning Milling Drilling
	Thermal-based methods	Laser machining Plasma arc machining Ion beam machining Electron beam machining
	Abrasive-based methods	Sandblasting Shot peening Shot blasting Water jet machining Magnetic abrasive machining
	Chemical methods	Chemical machining Electrochemical machining Plasma electrolytic machining
	Other methods	Electrical discharge machining Hybrid methods
Surface protection, functionalization, and decorative	Conventional method	Spraying Sandblasting
	Surface cleaning	Lye cleaning Solvent cleaning Chemical cleaning Mechanical cleaning
	Surface coating	Spraying Chemical oxidation Electroplating Surface alloying Physical vapor deposition Chemical vapor deposition Ion deposition oxide High-temperature oxidation
	Other methods	Electrochemical protection Temporary rust protection

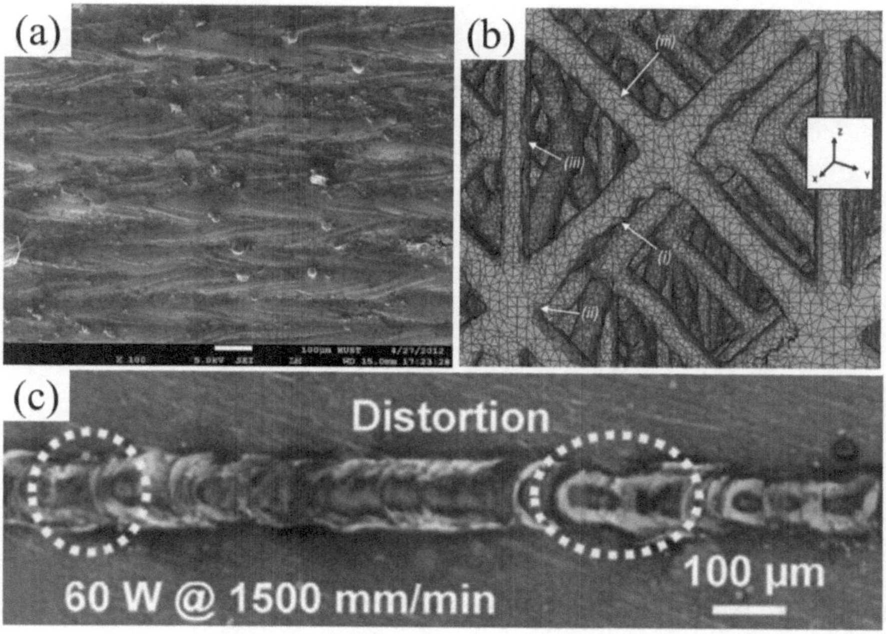

FIGURE 2.4 (a) Microscopic morphology characteristics and (b and c) distortion of as-built laser powder bed fusion (L-PBF) Inconel 625 samples (distortion in (b) the lattice structure and (c) the single track) (Tian et al. 2020).

2.5.1 Non-Thermal Treatment

Methods such as shot peening, spraying, and other deposition methods listed in Table 2.2 can be employed to refill porous structures, cracks and improve fatigue strength. Figure 2.5 shows the fatigue test result of SLM Ti6Al4V alloy after different surface treatment technologies. The use of shot peening and laser peening improves fatigue strength owing to the modification of the surface roughness and generation of a compressive residual stress field close to the surface (Aguado-Montero et al. 2022). Noticeably, shot peening is different from shot blasting although they have similar operation styles and belong to abrasive-based machining, as shown in Table 2.2. Shot peening is a cold work process used to impart compressive residual stresses onto the sample surface, aiming at improving mechanical properties. The cleaning materials removal efficiency is low and therefore is not recommended for polishing rough surfaces. The effect of shot peening for polishing AlSi10Mg samples is shown in Table 2.3. The polishing effect was not as good as conventional machining while the shot-peened surface gave the highest micro-hardness, as shown in Figure 2.6 (Maamoun, Elbestawi, and Veldhuis 2018). Shot blasting is a mechanical cleaning process that uses spheres of abrasive materials to remove oxides and debris, controlling the surface texture.

However, interior defects may require heat treatment since surface modification techniques have limited access to the interior structure. In addition, components may

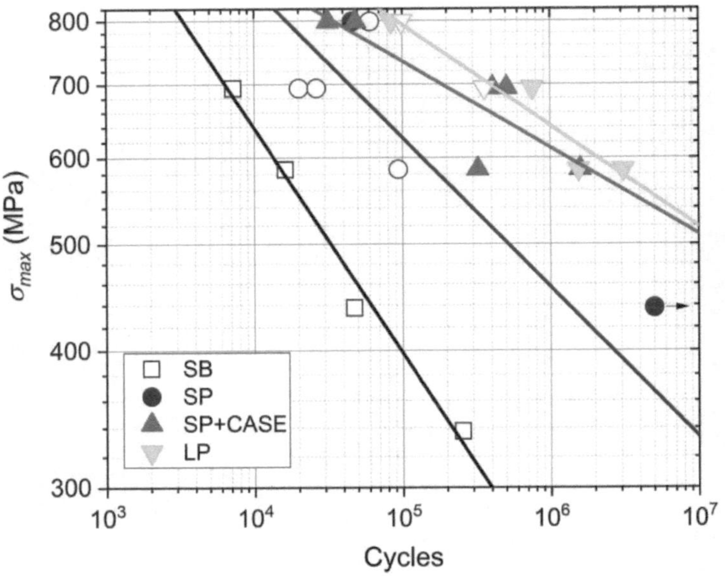

FIGURE 2.5 Fatigue test result of SLM Ti6Al4V (R = 0.1). Symbols are shown as solid for failures initiated from the interior and hollow for failures initiated from the surface (SB: sandblasting, SP: shot peening, SP+CASE: the combination of shot peening plus chemical assisted surface enhancement, LP: laser peening) (Aguado-Montero et al. 2022).

TABLE 2.3
The Effect of Shot Peening on Both Surface Roughness and Waviness (AB – As Built, SP – Shot Penning, M – Machine, HSP – High-Intensity SP, LSP – Low-Intensity SP) (Maamoun, Elbestawi, and Veldhuis 2018)

Surface	AB (Strip)	AB (Up-Skin)	AB + SP (Strip)	AB + SP (Up-Skin)	M	M + HSP	M + LSP
Average Ra (μm)	11.96	6.98	5.92	5.82	0.22	5.34	2.05
Surface waviness (μm)	0.38	2.2	2.79	1.92	0.06	0.23	0.07

not be fully cured after manufacturing by vat photopolymerization technologies, causing deterioration in hardness, ductility, etc. However, this can be eliminated by exposing the products to additional UV or visible light to re-cure the products.

2.5.2 Thermal Treatment

Thermal technologies are based on temperature control and sometimes combined with other parameters such as pressure, atmospheric composition, heating, and cooling rate. This type of technology can improve the performance of components by changing the microstructure inside the workpiece or the chemical composition of the

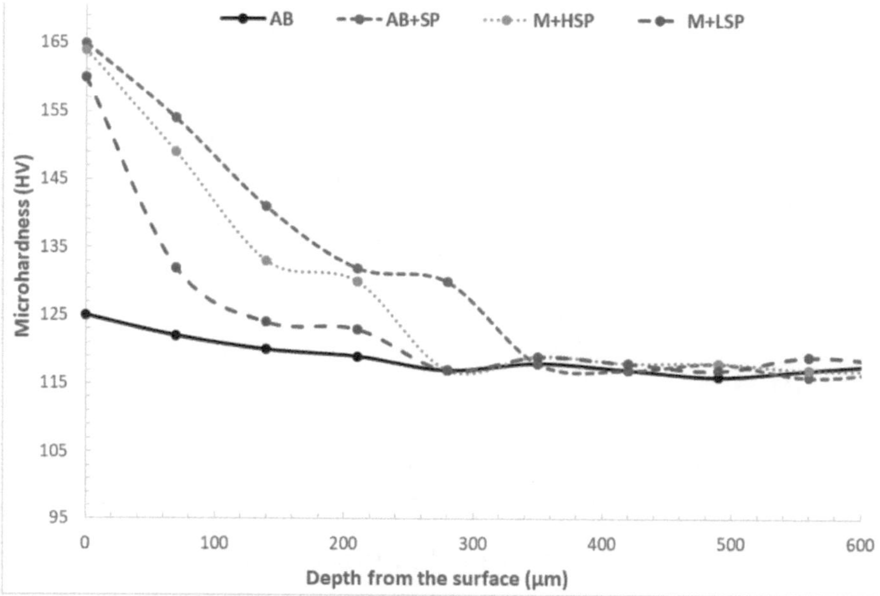

FIGURE 2.6 The in-depth microhardness profile of the AlSi10Mg samples under various SP conditions (Maamoun, Elbestawi, and Veldhuis 2018).

surface without affecting the components' geometry and the chemical composition of the bulk materials.

Components manufactured by powder-based AM processes usually have many pores inside or on the surface. Stress and deformation caused by non-uniform thermal distribution during the manufacturing process are also inevitable. These will severely deteriorate the fatigue life, hardness, ductility, etc. Annealing, tempering, etc., can be applied to improve mechanical properties by controlling the size, distribution of grains, and precipitations and phase transition. For example, the precipitation size of aluminium alloys grew from 12.0 to 20.0 nm with heat treatment under 325 °C from 4 to 48 hours. Since the region free of precipitates disappeared after 4 hours of heat treatment, the yield stress increased to 455.8 from 286.9 MPa of as-built state, and is proportional to the ratio of precipitation volume fraction to precipitation size, as shown in Figure 2.7 (Kuo et al. 2021).

Products made of polymers usually have lower mechanical properties and experience greater deformation at elevated temperatures. In this case, mold or powder can be introduced as 'support' to contain the polymer products and avoid deformation, such as thermal expansion or delamination of layers. For example, NaCl powder can be introduced as support for AM polyethylene terephthalate glycol polymer, as shown in Figure 2.8. After thermal processing at the temperature of 220 °C for 5 to 15 min, the samples displayed less localized necking and higher materials' diffusion. The tensile and compressive strength increased by 143% and 50%, respectively (Amza et al. 2021).

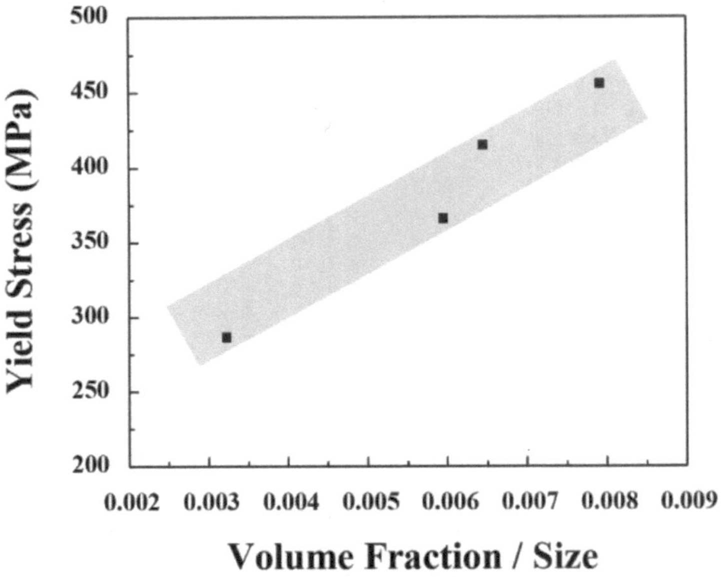

FIGURE 2.7 Schematic illustrations of the relation between the yield stress and ratio of precipitation volume fraction to size for the 3D-printed Scalmalloy (Kuo et al. 2021).

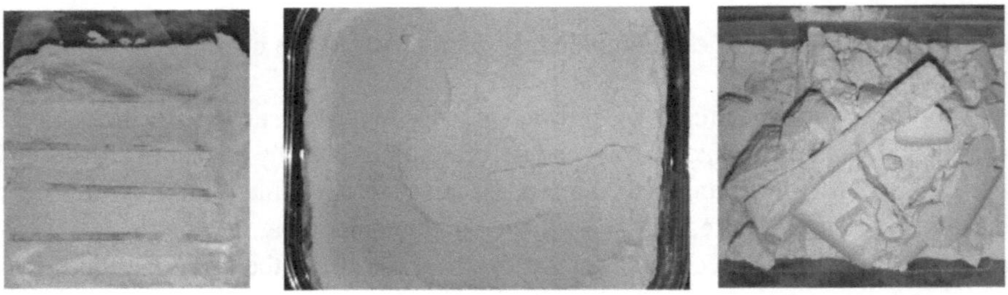

FIGURE 2.8 Heat treatment: (a) samples being powder-packed in the borosilicate glass recipient; (b) recipient with packed powder; (c) 3D-printed samples after heat treatment (Amza et al. 2021).

2.6 SUMMARY

This chapter gives an overview of the general post-processing technologies and procedures to provide an initial understanding of what and why post-processing is required. The majority of components fabricated using AM technologies may require one or more post-treatment steps to meet geometric, dimensional, and functional requirements. Common procedures can be summarized as primary cleaning, support structure removal, improvement in mechanical properties, secondary surface finishing, and surface coating or dyeing. In practice, however, the post-processing procedures should be determined by AM technologies, materials, and purposes, which requires practitioners to understand both AM technologies and the most appropriate post-processing technologies deployed.

REFERENCES

Aguado-Montero, Santiago, Carlos Navarro, Jesús Vázquez, Fernando Lasagni, Sebastian Slawik, and Jaime Domínguez. 2022. 'Fatigue Behaviour of PBF Additive Manufactured TI6AL4V Alloy after Shot and Laser Peening'. *International Journal of Fatigue* 154 (January): 106536. https://doi.org/10.1016/j.ijfatigue.2021.106536.

Amza, Catalin Gheorghe, Aurelian Zapciu, George Constantin, Florin Baciu, and Mihai Ion Vasile. 2021. 'Enhancing Mechanical Properties of Polymer 3D Printed Parts'. *Polymers* 13 (4): 562. https://doi.org/10.3390/polym13040562.

Dixit, Nitin, Varun Sharma, and Pradeep Kumar. 2021. 'Research Trends in Abrasive Flow Machining: A Systematic Review'. *Journal of Manufacturing Processes* 64 (April): 1434–61. https://doi.org/10.1016/j.jmapro.2021.03.009.

Gibson, Ian, David Rosen, Brent Stucker, and Mahyar Khorasani. 2021. *Additive Manufacturing Technologies*. Cham: Springer International Publishing. https://doi.org/10.1007/978-3-030-56127-7.

Gupta, Kapil, and Alokesh Pramanik. 2021. *Advanced Machining and Finishing*. 1st ed. Elsevier.

Jiang, Jingchao, Xun Xu, and Jonathan Stringer. 2018. 'Support Structures for Additive Manufacturing: A Review'. *Journal of Manufacturing and Materials Processing* 2 (4): 64. https://doi.org/10.3390/jmmp2040064.

Kuo, C. N., P. C. Peng, D. H. Liu, and C. Y. Chao. 2021. 'Microstructure Evolution and Mechanical Property Response of 3D-Printed Scalmalloy with Different Heat-Treatment Times at 325 °C'. *Metals* 11 (4): 555. https://doi.org/10.3390/met11040555.

Lou, Shan, Jane Jiang, Wenjuan Sun, Wenhan Zeng, Luca Pagani, and Paul Scott. 2019. 'Characterisation Methods for Powder Bed Fusion Processed Surface Topography'. *Precision Engineering* 57 (May): 1–15. https://doi.org/10.1016/j.precisioneng.2018.09.007.

Maamoun, Ahmed H., Mohamed A. Elbestawi, and Stephen C. Veldhuis. 2018. 'Influence of Shot Peening on AlSi10Mg Parts Fabricated by Additive Manufacturing'. *Journal of Manufacturing and Materials Processing* 2 (3): 40. https://doi.org/10.3390/jmmp2030040.

Maleki, Erfan, Sara Bagherifard, Michele Bandini, and Mario Guagliano. 2021. 'Surface Post-Treatments for Metal Additive Manufacturing: Progress, Challenges, and Opportunities'. *Additive Manufacturing* 37: 101619. https://doi.org/10.1016/j.addma.2020.101619.

Ni, Chenbing, Lida Zhu, Zhongpeng Zheng, Jiayi Zhang, Yun Yang, Jin Yang, Yuchao Bai, Can Weng, Wenfeng Lu, and Hao Wang. 2020. 'Effect of Material Anisotropy on Ultra-Precision Machining of Ti-6Al-4V Alloy Fabricated by Selective Laser Melting'. *Journal of Alloys and Compounds* 848 (December): 156457. https://doi.org/10.1016/j.jallcom.2020.156457.

Ribeiro, Kandice Suane Barros, Fabio Mariani, and Reginaldo Teixeira Coelhos. 2020. 'A Study of Different Deposition Strategies in Direct Energy Deposition (DED) Processes'. *Procedia Manufacturing*, 48th SME North American Manufacturing Research Conference, NAMRC 48, 48 (January): 663–70. https://doi.org/10.1016/j.promfg.2020.05.158.

Rosa, Benoit, Pascal Mognol, and Jean-yves Hascoët. 2015. 'Laser Polishing of Additive Laser Manufacturing Surfaces'. *Journal of Laser Applications* 27 (S2): S29102. https://doi.org/10.2351/1.4906385.

Sanaei, Niloofar, and Ali Fatemi. 2021. 'Defects in Additive Manufactured Metals and Their Effect on Fatigue Performance: A State-of-the-Art Review'. *Progress in Materials Science* 117 (April): 100724. https://doi.org/10.1016/j.pmatsci.2020.100724.

Santos, Ericles Otávio, Pedro Lima Emmerich Oliveira, Thaís Pereira de Mello, André Luis Souza dos Santos, Carlos Nelson Elias, Sung-Hwan Choi, and Amanda Cunha Regal de Castro. 2022. 'Surface Characteristics and Microbiological Analysis of a Vat-Photopolymerization Additive-Manufacturing Dental Resin'. *Materials* 15 (2): 425. https://doi.org/10.3390/ma15020425.

Shanbhag, Gitanjali, Evan Wheat, Shawn Moylan, and Mihaela Vlasea. 2021. 'Effect of Specimen Geometry and Orientation on Tensile Properties of Ti-6Al-4V Manufactured by Electron Beam Powder Bed Fusion'. *Additive Manufacturing* 48 (December): 102366. https://doi.org/10.1016/j.addma.2021.102366.

Smith, Peter H., James W. Murray, Daniel O. Jones, Joel Segal, and Adam T. Clare. 2021. 'Magnetically Assisted Directed Energy Deposition'. *Journal of Materials Processing Technology* 288 (February): 116892. https://doi.org/10.1016/j.jmatprotec.2020.116892.

Thomas, Douglas S., and Stanley W. Gilbert. 2014. 'Costs and Cost Effectiveness of Additive Manufacturing'. *NIST*, December. https://www.nist.gov/publications/costs-and-cost-effectiveness-additive-manufacturing.

Tian, Zhihua, Chaoqun Zhang, Dayong Wang, Wen Liu, Xiaoying Fang, Daniel Wellmann, Yongtao Zhao, and Yingtao Tian. 2020. 'A Review on Laser Powder Bed Fusion of Inconel 625 Nickel-Based Alloy'. *Applied Sciences* 10 (1): 81. https://doi.org/10.3390/app10010081.

Unkovskiy, Alexey, Phan Hai-Binh Bui, Christine Schille, Juergen Geis-Gerstorfer, Fabian Huettig, and Sebastian Spintzyk. 2018. 'Objects Build Orientation, Positioning, and Curing Influence Dimensional Accuracy and Flexural Properties of Stereolithographically Printed Resin'. *Dental Materials* 34 (12): e324–33. https://doi.org/10.1016/j.dental.2018.09.011.

Zhang, Jun, Negin Amini, David A. V. Morton, and Karen P. Hapgood. 2021. 'Binder Jetting of Well-Controlled Powder Agglomerates for Breakage Studies'. *Advanced Powder Technology* 32 (1): 19–29. https://doi.org/10.1016/j.apt.2020.11.012.

Zhu, Haitao, Allan Rennie, Ruifeng Li, and Yingtao Tian. 2022. 'Two-Steps Electrochemical Polishing of Laser Powder Bed Fusion 316l Stainless Steel'. *Surfaces and Interfaces* 35 (December): 102442. https://doi.org/10.1016/j.surfin.2022.102442.

3 Economic and Environmental Aspects of Post-Processing Technologies in Additive Manufacturing

Balwant Singh, Jasgurpreet Singh Chohan, and Raman Kumar
Department of Mechanical Engineering, Chandigarh University, Gharuan, Mohali, Punjab, India

CONTENTS

3.1 INTRODUCTION

Unlike traditional manufacturing methods, additive manufacturing (AM), often known as three-dimensional (3D) printing, is the process of combining materials from 3D model data to create layered objects on a layer. This tool-free production strategy can enable freedom in design, personal customization, high precision in complicated products, decrease energy and material use, and shorten time to market

DOI: 10.1201/9781003276111-3

by eliminating the need for additional equipment like tools, gauges, or fixtures. One of the disadvantages of AM technologies, on the other hand, is that they yield components with poor surface quality. Surface roughness, dimensional consistency, and structural supports are all surface-related issues that prohibit parts from becoming finished products. Given all of this, it is clear that post-processing operations are required to obtain ready-to-use components. Chemical post-processing, mechanical post-processing activities, and thermal post-processing operations are the three types of post-processing processes (Jiménez, Romero, Domínguez, Espinosa, & Domínguez, 2019). Specific application and material, each has its own set of benefits and drawbacks.

Post-treatment is often necessary after the completion of an additive manufacturing (AM) part to separate the parts from the build plate, remove support structures, or get the required dimensions and/or surface characteristics. Electrical discharge machining (EDM), ultrasonic cleaning, and conventional machining methods are all common AM post-treatment processes. Chemical post-processing procedures are popular because they improve the surface quality of the parts while also being cost-effective and simple to utilize. Chemical post-processing procedures are commonly utilized in the biomedical industry for surface cleaning and polishing of parts. Chemical post-processing procedures are commonly employed on ABS (acrylonitrile-butadiene-styrene) parts created with FDM (fused deposition modeling) parts, according to the literature, because chemical solutions permeate the plastic parts better. Mechanical post-processing procedures enhance part quality and ensure that they fulfill design criteria. Surface qualities, geometric accuracy, aesthetics, mechanical properties, and more can all be improved through mechanical post-processing techniques. The thermal post-processing approach for AM parts can greatly minimize residual stresses, fracturing, and microstructure homogenization (Peng, Kong, Fuh, & Wang, 2021). The types of post-processing techniques have been shown in the following Figure 3.1.

All of these post-processing techniques have some aspects towards environmental and economic factors. Environmental aspects mean the operations, items, or services that have the potential to interact with the environment (Khalid & Peng, 2021). Activities related to any of these components are carefully

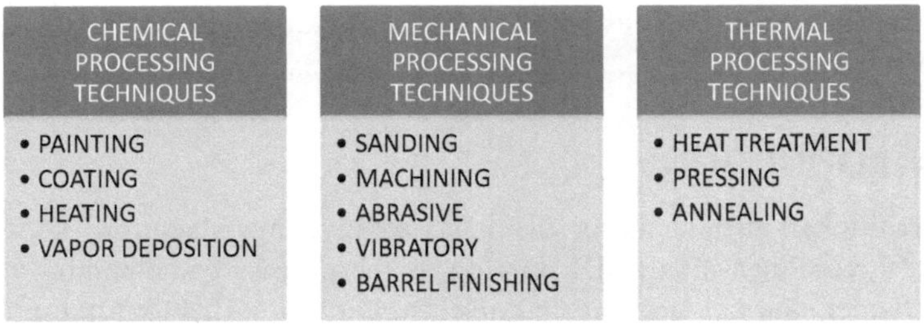

FIGURE 3.1 Post-processing techniques.

monitored to reduce or eliminate environmental impacts. The following factors have been highlighted as having the potential to have a substantial impact on the environment:

(i) noxious waste;
(ii) waste-containing radioactive elements;
(iii) waste that has been mixed;
(iv) emissions allowed in the air;
(v) liquid discharges that are regulated;
(vi) chemical storage, use, and transportation;
(vii) radioactive materials: storage, use, and transportation;
(viii) non-renewable energy intensity/usage (electricity);
(ix) emissions of greenhouse gases.

Each action has a result. Some may have relatively minor environmental effects, such as fewer waste and reduced energy consumption, both of which contribute to air pollution, like those offered by office-based services. While some heavy industrial processes, like those that cause air emissions and water discharges, could have negative environmental effects. Managing environmental factors and effects is the most important aspect of an environmental management system.

The economics of post-processing of additive manufacturing is divided into three categories. The first involves measuring the size of the post-processing process of additive manufacturing. This includes calculating the market value of goods produced with this technology about the overall economy. The second step is to calculate the costs and advantages of implementing this technology. It includes knowing when and why post-processing methods of additive manufacturing are more cost-effective than traditional manufacturing. It also entails comprehending other benefits, such as the ability to create new items that would otherwise be impossible with traditional manufacturing (Bakker, Whan, Knap, & Schmeits, 2019).

3.2 POST-PROCESSING TECHNIQUES

After the additive manufacturing process, post-processing techniques are used to maintain the high quality of the product. Post-processing is the final step in the production process, where items are smoothed and painted, among other things. Parts undergo post-processing to improve their quality and guarantee that they fulfill design criteria. Surface qualities, geometric accuracy, aesthetics, mechanical properties, and more can all be improved through the finishing process. This can imply the difference between a sale and a loss for samples and prototypes. Finishing generates a ready-to-use portion for production parts. When making an order for additives, it's critical to consider the finishing capability of the additive manufacturer. Some businesses devote only a tiny portion of their space to completing. This can result in high-cost

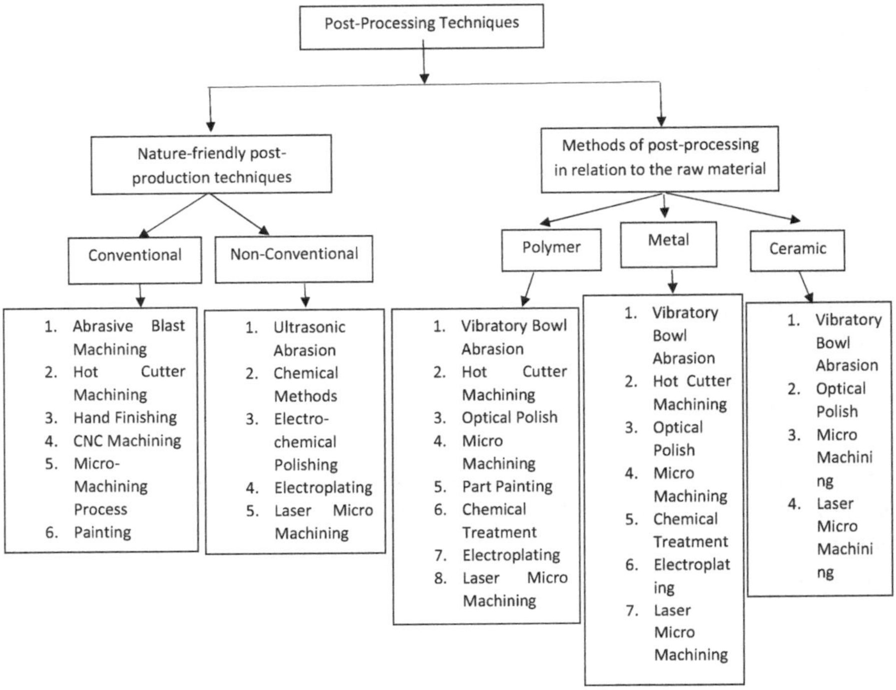

FIGURE 3.2 Methods of post-processing technique on the basis of conventional and non-conventional processes.

post-processing, outsourcing, or bottlenecks during the production's final stages (Mahmood, Chioibasu, Rehman, Mihai, & Popescu, 2022). It might also serve as a motivator to complete the task quickly. Based on conventional and non-conventional procedures, various post-processing techniques are listed in Figure 3.2.

3.3 ENVIRONMENTAL ASPECTS

The misuse and destruction of our natural planet have been a result of post-processing technology's impact on the environment. When dangerous or excessive amounts of gases, such as carbon dioxide, carbon monoxide, sulfur dioxide, nitric oxide, and methane, are injected into the Earth's atmosphere, it is known as air pollution. The main sources all deal with technology that came after the industrial revolution, such as fossil fuel combustion, factories, power plants, mass agriculture, and automobiles. The effects of air pollution include poor health repercussions for humans and animals and global warming, whereby the increasing amount of greenhouse gases in the air traps thermal energy in the Earth's atmosphere and cause the global temperature to rise. As a result of human activities, water pollution contaminates bodies of water like reservoirs, canals, oceans, and groundwater. The most common water contaminants include household waste, industrial runoff, insecticides, and pesticides. Aquatic ecosystems may be destroyed, for example, as a result of the discharge of wastewater that has not been treated properly into natural reservoirs. Other detrimental effects on the food supply chain include cholera and typhoid outbreaks, eutrophication, and habitat damage. Resource depletion is another negative impact of technology on the environment. When a resource is used up more quickly than it can be replenished, it happens. Natural resources can be renewable or non-renewable and are those that already exist in the world without being produced by humans. The most severe types of resource depletion are aquifer depletion, deforestation, mining for minerals and fossil fuels, resource pollution, soil erosion, and overconsumption. The main causes of agriculture, mining, water use, and fossil fuel consumption have all been made possible by technical advancements. The rate at which natural resources are being degraded increases together with the world population. As a result, it's estimated that the world's eco-footprint is 1.5 times more than the capacity of the planet to sustainably supply each person with enough resources to match their consumption levels. Since the industrial revolution, there has been an increase in large-scale mineral and oil exploration, which has led to a growing natural resource depletion. Technology, research, and development advancements have made it simpler to mine minerals, and people are digging deeper to obtain more of them, which has led to a decline in the production of numerous resources. Despite the negative effects of technology on the environment, a recent rise in concern about climate change has prompted the development of new environmental technology. This technology is intended to help society transition to a more sustainable, low-carbon economy, which will help resolve some of the most pressing environmental issues we face. The development of new technologies to protect, monitor, or minimize technology's negative effects on the environment and resource consumption is referred to as environmental technology, sometimes known as "green" or "clean" technology (Gibert et al., 2008).

Every post-processing technique has its aspects toward the environment. A detailed discussion about the different post-processing techniques in the environment has been discussed below:

3.3.1 Impact of Abrasive Blast Machining on the Environment

Abrasive substances break down into incredibly small dust particles when they come into contact with a surface substance. Due to the fact that these particle emissions are a substantial source of air pollution, abrasive blasting and the environment are unlikely to get along. Depending on blast factors like the type of equipment used, the blast velocity, the blast angular velocity, the range from the component being blown, and the part measurements, particle emissions from blasting frequently contain varying concentrations of silica, aluminum, arsenic, copper, and other hazardous materials. The effects on the ecosystem can be catastrophic when these tiny emissions of dust contaminate the air. Studies show that these pollutants can harm the environment in such a way that they can modify weather patterns, bring on droughts, contribute to climate change, and even cause ocean acidification. Particle emissions into the atmosphere enable the warming effect, which traps the air's heat and prevents it from escaping.

A water having high velocity jet with abrasive particles is used in water jet cutting to remove a wide range of materials. Serious physical harm will happen from direct interaction with the jet, such as removing a leg from a careless operator. The jet user must maintain a secure distance from it and must adequately cover it. Eye injury is a frequent worry with water jet cutting. Consequently, eye safety is essential. When a water jet cuts, a lot of noise is made, and if the user is subjected to it over an extended period of time, it might harm their hearing. Regular hearing tests are required for water jet cutters. Another major issue with abrasive jet machining (AJM) is the creation of silica dust when the abrasive substrate or material properties comprises silica. The operators can breathe in the particles if they are small enough, and they have the potential to harm the environment. A disorder called silicosis, which makes the lungs harden and scar, can be brought on by silica dust. Chest pain, coughing, and shortness of breath are silicosis symptoms. Lung cancer risk can increase as a result of prolonged silica exposure. To reduce the danger of silica poisoning, a variety of prevention methods can be implemented, such as reducing or eliminating its use, if possible, isolating the process, or putting in place proper administrative procedures. The components of the workpiece employed in this process could potentially injure the operators. The risk factors for breathing in hazardous metal dust depending on the type, shape, intensity, biologic impact, and exposure duration to the particles. It is possible that the particles are too small to be absorbed and can only be gathered by the respiratory system, or they may be so massive so as to prevent inhalation and collection by the respiratory system. The skin may become pierced by tiny metal or abrasive particles, which could result in pain or illness. To avoid coming into contact with the particles, the workstation must be kept tidy. Use the proper respiratory protection gear to protect yourself against harmful atmospheres. It is vital to dust and clean correctly. Soaking up dust before cleaning is one cleaning technique. The workpiece could become embedded with the cutting abrasives, lowering the quality of the finished output. The sand left over after water jet cutting can make the shop floor dusty, which could affect the adjacent machines. To enhance working conditions, general control procedures including installing effective aeration, an air filtering system, and an appropriate dust storage system. Because of the heavy physical and mental exhaustion, low job satisfaction, and high error rates caused by the dust, the operators' working environment is unpleasant, which indirectly reduces productivity. Additionally, operators must use safety tools and wear personal protective equipment (PPE) (Zulkarnain et al., 2021).

3.3.2 Controlling Techniques used in Abrasive Jet Machining

FOR BETTER ENVIRONMENT: Strict pollution regulations and dust suppression procedures are necessary due to the major environmental risks connected to abrasive blasting in order to guarantee that the emissions produced during blasting are captured and recovered (Anu Kuttan, Rajesh, & Dev Anand, 2021).

(i) ENCLOSURES FOR BLAST: In order to keep particulate emissions from entering the atmosphere, blast enclosures are made to completely enclose abrasive blast activities in a basket framework. Many blast enclosure versions employ ventilation systems to remove the dust from the cage and effectively eliminate the dirt from the atmosphere.

With just trace amounts of waste leaking out at the blast container seams, blast enclosures are frequently highly effective at containing and collecting abrasive blasting particulate pollution. Blast enclosures can be expensive, and they typically make cleaning take longer (Kumar, 2019).

(ii) BLASTERS WITH A VACUUM: For cleaning the carpet and floors in your home, vacuum blasters function similarly to a vacuum cleaner. A catch and collection device that stores the emissions surrounds the blast nozzle. The collected abrasives can then be used once more during the blasting process.

Although this technique is effective for collecting particle emissions, it is similar to blast enclosures in that it is expensive and challenging to locate. Vacuum blaster systems can also be cumbersome, heavy, and challenging to use (Benedict, 2018).

(iii) DRAPES: Drapery, often known as curtains, is a crucial sort of particle management that aids in ensuring safety to both individuals and the environment. Drapes with perforations are hung on either side of a truss-like structure to direct waste into a lined net that is put beneath the blasting operations to catch any waste generated during the blasting process.

Not usually the best method for particle control is draperies. In rare instances, dust may seep through the drapes, and additional dust leakage is frequently caused by wind or gusts near the explosion zone. However, curtains are typically one of the most affordable control alternatives while still offering vital security for both the ecosystem and the workforce.

(iv) WATER CURTAINS: A number of nozzles must be placed along the perimeter of the surface or building being blasted in order to use the regulation technique known as water curtains. The trash from the blasting operation is directed and collected by a water curtain created by these nozzles.

Following that, this floating garbage is transported to the ground, where it gathers in water troughs. The abrasive substance mixture can then be discarded in the appropriate container.

A low-cost particle collection technique that is effective at reducing hazardous particle discharges during blasting is the use of water curtains.

(v) WET BLASTING: Wet abrasive blasting, high-pressure water blasting, high-pressure water and abrasive blasting, and air and water abrasive blasting are all examples of wet blasting.

This control approach works by mixing water and abrasive particles in the abrasive blasting nozzle. This mixture catches dust emissions as soon as they hit the ground, preventing any damage to the environment or personnel.

Even though wet blasting works at gathering dust, it may take longer to clean the surface than dry blasting because it uses water. You can fix this issue by covering the blast nozzle's end with a retrofit device.

(vi) CENTRIFUGAL BLASTERS: In centrifugal blasters, abrasives are forced against the surface of whatever is being cleaned by a fast-moving blade. After that, the abrasive is recycled using a collection system that almost eliminates debris escape.

Although efforts have been made to build a smaller, hand-held variation for more versatile application, usually huge horizontal constructions like ship decks call for the usage of these blasters. Use these portable centrifugal blasters on bridges and other related constructions (Rao, Naidu, & Kona, 2018).

3.3.3 Impact of Chemical Machining on the Environment

Chemical machining (CHM) is bad for the environment since it uses and discards dangerous chemicals like etchants, cleaners, and strippers. Additionally, their handling and disposal incur a great deal of expense. Another effect of the chemicals that harms many materials is acidic corrosion. The soil and water's acidity and alkalinity levels are changed by the etchants used in CHM, which has an impact on the flora and wildlife. Aquatic life experiences difficulties in surviving whenever the pH of ordinary water changes as a result of pollution. Aerosols of liquid or solid corrosive chemicals (such as nitrogen and sulfuric oxides) can be used in chemical machining to create corrosive gases. These gases may pollute the environment, especially the soil and water, because they can produce acid when combined with water. Chemical machining frequently has negative health impact, include irritability, caustic cuts and blisters, quick, serious, and sometimes permanent vision problems, larynx carcinoma, and emphysema. The characteristics, concentration, and duration of the substances' exposure to acids and alkali are what determine any potential health risks (El-Hofy & Youssef, 2009).

3.3.4 Impact of Electrochemical Machining on the Environment

Environmental pollution is caused by both the electrolyte and the slurry used in electrochemical machining (ECM). The environment is not harmed by the fresh electrolytes in ECM. On the other hand, as heavy metals from the components of the workpiece build up, the electrolyte becomes contaminated over time. Chemical by-products including poisonous chromate, nitrate, and ammonia, as well as others, can attach in the EC sludge to iron oxides. These might have the capacity to contaminate the land, water, or air. Economic effectiveness of ECM is constrained by the costs of detoxicating and disposing of hydroxide slurries. The electrolysis process can result in the explosive production of hydrogen gas. Human skin, liver, and kidneys can become damaged by chromium compounds by ingestion, inhalation, contact with the eyes or skin, and contact dermatitis. Process enclosures, local exhaust ventilation, general dilution ventilation, and personal protective equipment are other safety and health management strategies that may be helpful.

3.3.5 Impact of Electro-Discharge Machining

In electro-discharge machining (EDM), material is removed through a series of sparks that form in the dielectric fluid that fills the space between the tool-electrode and the workpiece. EDM's ability to accurately cut robust materials has contributed to its rise in popularity. Hazardous emissions, such as benzene, mineral vapors, are created while using the EDM as a result of the usage of organic dielectric liquids or oil products. Other consequences of the typical utilizing liquid solutions in wire EDM include toxic aerosols, carbon monoxide, nitrous oxide, and ozone. The exhaust from electro-discharge machining is influenced by the metal removal rate method, the dielectric, and the workpiece material. When dielectric is inhaled under the guise of smoke, vapors, or aerosols, most people have negative health effects. Dielectric vaporization can release organic molecules that can pollute the environment. In addition, the deterioration of tool and workpiece electrodes results in the production of inorganic substances, including carbide-coated tungsten and carbide-coated titanium, besides condensation in the air that pollutes the environment. Dielectric waste, deionizing resin, and EDM sludge materials extracted from both the workpiece and the tool have negative effects on the environment if they are not disposed of appropriately. Several preventive measures must be addressed to reduce any possible dangers associated with the EDM machining process. To decrease, a secure exhaust is required of fume or breathing in smoke. The effective use of filters and the disposal of garbage can both reduce environmental pollution. The medium's temperature must be kept below the flashing point to prevent a fire danger. A maintenance schedule should include checking the dielectric level and cleaning the dielectric. If the EDM machine is adequately insulated, the amount of electromagnetic radiation it produces can be decreased. In order to prevent severe harm or even death, operators must be made aware of the risks associated with high voltage. There is a rising need for a less hazardous fluid because the dielectric fluid is largely to blame for health and environmental issues.

3.3.6 Impact of Laser Beam Machining

In laser machining, a powerful light beam is used to heat, partially evaporate, and modify the material to be sliced chemically. The production of aerosols and gases, as well as radiation exposure from high energies, are the primary risks that laser cutting poses. The workpiece material is vaporized, and a variety of particle sizes are created. Workers who use laser machining may be at risk for health problems if they are exposed to nanoparticles. According to toxicological study, the concentration and particular surface area of the particles affect the negative consequences on one's health of breathing in nanoparticles. Due to the release of smells and air pollutants during laser cutting of polymers and conducting polymers, this procedure is also considered as hazardous. According to laboratory tests using laser cutting to heat or vaporize polymers, widely used plastics can emit a range of airborne contaminants, including carcinogens and respiratory irritants. Depending on the type of plastic, formulation, and processing conditions, the vapors created when plastic is heated can have a complex composition. Using a lot of assisting gases, such CO_2, could be hazardous.

3.3.7 Impact of Electron Beam Machining

In order to remove material, electron beam machining (EBM) accelerates electrons toward the workpiece at speeds between 50 and 80 percent of light using a high voltage (often 120 kV). Even though EBM is a clean machining method, if considerable

X-ray leakage happens, the operator's health could be at risk due to the workpiece's and the laser beam's contact. To avoid any possible health issues, think about building a proper box, using lead-based glass in the viewing ports, adding a locking door, following the right procedure, and ensuring trained employees run the apparatus. The production of dust as a result of the melting of the metal during EBM is another legitimate worry. Metal dust accumulates on the room's interior walls. When positioning the workpiece before cutting and removing it after machining, the operator will be exposed to dust. High voltages connected to the electron gun and leftover heat in the workpiece could be hazardous to your health. Dust exposure can be reduced by using a dust control strategy and putting on personal protective equipment.

3.3.8 Impact of Ultrasonic Machining

Ultrasonic machining (USM) is a process for removing hard and brittle materials from an abrasive slurry. It involves an axially rotating tool and an ultrasonic frequency (18–20 kHz). The USM process can be used to cut glass, ceramics, and materials made of carbide. While utilizing this process, there are possible risks to both human health and the environment from abrasive slurry, electromagnetic fields, and ultrasonic waves. Although ultrasonography in USM produces an inaudible high-frequency noise, it might nonetheless cause damage to your ears. Furthermore, it has been discovered that ultrasonic frequencies can produce sounds that are audible and in the 96–105 dB range, which can be damaging to hearing. Other side effects of noise include dizziness, exhaustion, tension, pain, and hearing loss. Furthermore, noise may make it more difficult to communicate, which may lead to accidents. If necessary, PPE like earplugs must be worn. Noise can be muffled at the source by building an enclosure out of acoustic insulation material. Thus, long-term hearing damage will be avoided. Cutting fluids used in conventional machining have different effects on human health and the environment than abrasive slurry fluids used in USM. Skin disorders such as dermatitis could develop after coming into contact with the slurry. When breathed in, the slurry mist could be suspended in the air of the worker's breathing zone and cause respiratory conditions, like bronchitis and asthma. Engineering measures like installing a mist collector or adding anti-mist chemicals to the slurry fluid can significantly limit the amount of mist exposure. Long-term exposure to a high electromagnetic field (EMF) might be harmful to your health. As the EMF strength rapidly decreases with distance from the source, health hazards can be avoided by maintaining a safe distance from the source (Gamage & De Silva, 2015).

3.4 Economical Aspects

Post-processing is used for AM parts not only because present technologies and procedures are unable to make components with the appropriate net shapes, but also to achieve more favorable mechanical qualities that AM processes have yet to accomplish. Post-processing techniques aid in reducing research and development costs, transaction costs, new and improved product benefits, long-term growth and competitiveness benefits, and other societal benefits. Beyond the public good nature of technology infrastructure, it identifies that distinct hurdles to innovation driven by market failures. These impediments exacerbate inefficiency and highlight the need for public institutions in satisfying scientific and technological requirements. Because of infrastructural gaps, crucial uncertainties, and network effects and deployment is prohibitively expensive and only available to a few organizations.

End users are frequently unable to verify the quality and performance of post-processed parts, or even to determine whether potential issues with processed part performance and reliability are caused by underlying issues with post-processed manufactured parts, post-processing equipment, or even the powders and other raw materials, according to the analysis. End users cannot verify the quality of raw materials such as powders, and powder makers cannot verify the quality of private data and tests provided by them. The interviews show how proprietary standards impair markets by transferring market power through branding and reputation. As a result, post-processing research, development, and deployment are exorbitantly expensive, private investment incentives are undermined, and privately established standards may distort the market even more.

There are numerous interconnections between the various components of the technological infrastructure. This emphasizes the necessity of meeting demands in all categories. As a result, "unbalanced" investment in the domestic advanced manufacturing sector, which closes some technical gaps while leaving others unfulfilled, "would likely fail to fully realize the economic effect." According to the findings, present inadequacies in Post Processing technology infrastructure are particularly detrimental to small businesses. "They do not have adequate resources to extensively test and confirm part properties after fabrication and, as a result, are prevented from joining the transportation markets". Similarly, a minor materials provider said that the development cycles for new materials were too long to generate a profit (Godina et al., 2020). "Property data for a set of the common process–material couples could hasten the entry of additively manufactured components into service in existing as well as new businesses in this context, and open up further opportunities for small suppliers and manufacturers".

3.5 Future Scope and Conclusion

By describing several post-processing technologies and their applications in AM processes, including thermal post-processing, laser peening, laser polishing, machining, and abrasive finishing methods, this work has analyzed many techniques to improve the surface finish of the AM products. The term "post-processing" refers to a variety of steps that 3D-printed parts in the context of metal AM technology must go through before being used for their intended use, such as powder removal, stress relief annealing, wire cutting, other finishing, hot isostatic pressing, and so forth. Some of these processes still call for physical labor, where qualified operators are required for certain duties. Even though it could be more affordable to manually complete a prototype or a few dozen parts, the need for AM post-processing automation increases dramatically when thousands or even hundreds of components are produced.

To achieve automated post-processing, there are only a few centralized specialist solutions available, and these systems are mostly made for polymer AM parts. Automated solutions can increase production efficiency. The post-processing technology from conventional manufacturing is still applied to metal additive manufacturing. Some businesses have also started implementing robotic systems that can install printing substrates, clean powder, unload parts, and perform post-processing to further automate these technologies. To encourage continuous and massive manufacturing, it is intended to completely replace all manual labor. Even though this development is positive, the rate of innovation in this industry is still quite slow. To keep up with the AM industry's rapid expansion, there will undoubtedly be an increase in the number of sophisticated automatic post-processing systems in the future.

REFERENCES

Anu Kuttan, A., Rajesh, R., & Dev Anand, M. (2021). Abrasive water jet machining techniques and parameters: a state of the art, open issue challenges and research directions. *Journal of the Brazilian Society of Mechanical Sciences and Engineering*, *43*(4). https://doi.org/10.1007/s40430-021-02898-6

Bakker, K., Whan, K., Knap, W., & Schmeits, M. (2019). Comparison of statistical post-processing methods for probabilistic NWP forecasts of solar radiation. *Solar Energy*, *191*(April), 138–150. https://doi.org/10.1016/j.solener.2019.08.044

Benedict, G. F. (2018). Abrasive water jet machining (AWJM). *Nontraditional Manufacturing Processes*, (December), 37–51. https://doi.org/10.1201/9780203745410-4

El-Hofy, H., & Youssef, H. (2009). Environmental hazards of nontraditional machining. *IASME/WSEAS International Conference on ENERGY & ENVIRONMENT (EE'09)*, (August), 140–145.

Gamage, J. R., & De Silva, A. K. M. (2015). Assessment of research needs for sustainability of unconventional machining processes. *Procedia CIRP*, *26*, 385–390. https://doi.org/10.1016/j.procir.2014.07.096

Gibert, K., Izquierdo, J., Holmes, G., Athanasiadis, I., Comas, J., & Sànchez-Marrè, M. (2008). On the role of pre and post-processing in environmental data mining. *Proc. IEMSs 4th Biennial Meeting–Int. Congress on Environmental Modelling and Software: Integrating Sciences and Information Technology for Environmental Assessment and Decision Making, IEMSs 2008*, *3*, 1937–1958.

Godina, R., Ribeiro, I., Matos, F., Ferreira, B. T., Carvalho, H., & Peças, P. (2020). Impact assessment of additive manufacturing on sustainable business models in industry 4.0 context. *Sustainability (Switzerland)*, *12*(17), 0–21. https://doi.org/10.3390/su12177066

Jiménez, M., Romero, L., Domínguez, I. A., Espinosa, M. D. M., & Domínguez, M. (2019). Additive manufacturing technologies: An overview about 3D printing methods and future prospects. *Complexity*, *2019*. https://doi.org/10.1155/2019/9656938

Khalid, M., & Peng, Q. (2021). Sustainability and environmental impact of additive manufacturing: A literature review. *Computer-Aided Design and Applications*, *18*(6), 1210–1232. https://doi.org/10.14733/cadaps.2021.1210-1232

Kumar, R. (2019). *A Detailed Report on Process Parameters of Material Removal Rate in Abrasive Jet Machining*. (December).

Mahmood, M. A., Chioibasu, D., Rehman, A. U., Mihai, S., & Popescu, A. C. (2022). Post-processing techniques to enhance the quality of metallic parts produced by additive manufacturing. *Metals*, *12*(1). https://doi.org/10.3390/met12010077

Peng, X., Kong, L., Fuh, J. Y. H., & Wang, H. (2021). A review of post-processing technologies in additive manufacturing. *Journal of Manufacturing and Materials Processing*, *5*(2). https://doi.org/10.3390/jmmp5020038

Rao, P. S. V. R., Naidu, A. L., & Kona, S. (2018). Design and fabrication of abrasive jet machine (AJM). *Mechanics and Mechanical Engineering*, *22*(4), 1471–1482. https://doi.org/10.2478/mme-2018-0115

Zulkarnain, I., Mohamad Kassim, N. A., Syakir, M. I., Rahman, A. A., Md Yusuff, M. S., Yusop, R. M., & Keat, N. O. (2021). Sustainability-based characteristics of abrasives in blasting industry. *Sustainability (Switzerland)*, *13*(15), 1–13. https://doi.org/10.3390/su13158130

4 Post-Processing in Additive Manufacturing

Requirements, Theories, and Methods

Hamaid Mahmood Khan[1], *Gökhan Özer*[1], *and Mustafa Safa Yilmaz*[2]

[1]Fatih Sultan Mehmet Vakif University, Aluminium Test Training, and Research Center (ALUTEAM), TR 34445, Halic Campus, Istanbul, Turkey

[2]Bursa Uludag University, Engineering Faculty, Mechanical Engineering Department

CONTENTS

DOI: 10.1201/9781003276111-4

4.1 INTRODUCTION

3D printing has always been seen as a promising technology for creating small, low-quality prototypes. As it is getting faster and more dependable, its applications are becoming more diversified. 3D printing involves adding components layer by layer using only raw material and energy source, which is more economical and efficient with materials than cutting a shape from a larger block or pouring molten material into a mold [1–3]. In recent years, the commercial market for 3D printers has rapidly expanded thanks to an unprecedented development in materials and printing techniques, making it more diversified and highly competitive [3, 4]. With 3D printing's quick and simple design process, it is now possible to customize a wide variety of materials for different applications, making the development of unique products a reality that was previously only a pipe dream with milling or casting. The schematic representation of an LPBF operation is given in Figure 4.1.

Examples of unique designs include lattice materials, medical implants, and conformal cooling channels at an affordable cost, short production cycle, small batches, and more [2, 4, 5]. Additionally, the non-equilibrium solidification procedure might well be modified to provide materials with the desired characteristics for specified purposes. Furthermore, a wide variety of materials can be processed by optimizing energy density to the point where the great majority of materials can be locally heated to a high temperature [6].

4.1.1 AM-Based Terminologies

The commercial market of metal-based additive manufacturing (AM) methods is still dominated by powder bed fusion (PBF) and directed energy deposition (DED)

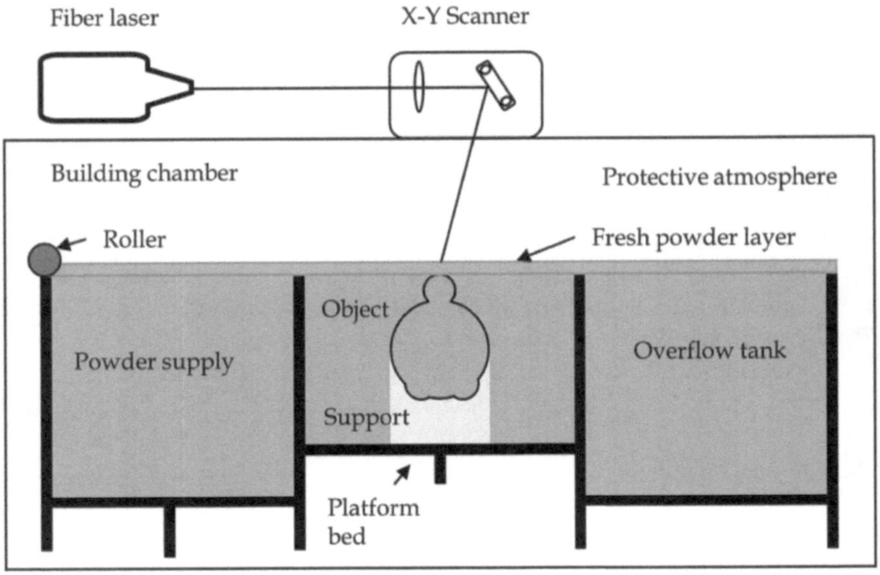

FIGURE 4.1 Schematic representation of a LPBF process.

methods. PBF method also supports polymers and, to a lesser extent, ceramics and composites [7]. Significant improvements to the AM laser induction and powder deposition system led to the discovery of the dedicated metal-based selective laser melting (SLM), direct metal laser sintering (DMLS), and electron beam melting (EBM) processes, which were initially designed as a manufacturing technique to produce plastic prototypes [8]. In the PBF system, discrete powder particles are fused in a layer in a protective environment of argon/nitrogen/vacuum using a computer-controlled heat source, such as a laser or electron beam, to reconstruct a two-dimensional (2D) contour of a three-dimensional (3D) computer-aided-design (CAD) model [7, 8].

In contrast to SLM, which employs a robust laser source to assemble powder particles in a nitrogen- or argon-rich environment, EBM uses an electron beam in vacuum to create solid components. DED, on the other hand, uses a nozzle to sequentially add energy and feedstock material—either metal powder or wire—to the build surface in an airtight chamber. The energy source for DED can be a laser, electron beam, or arc, depending on the version [8, 9]. In comparison to PBF processes, DED approaches can build large structures and fix damaged areas, though at a lesser resolution. Additionally, it moves forward quite quickly, consuming materials more effectively and minimizing expense. Contrary to PBF methods, DEDs are insufficient for creating complex shapes like cooling channels and lattices. Better than EBM, SLM can process a wide range of materials, including titanium, aluminum, steel, CoCr, Inconel, and other composites [9, 10]. In other words, additive manufacturing techniques allow designers to swiftly produce parts with intricate qualities, leading to major improvements in a component's fundamental design. The current literature will now frequently use these AM-based technologies to discuss how post-processing techniques affect them. Comparison of additive manufactured components with traditional ones is given in Figure 4.2.

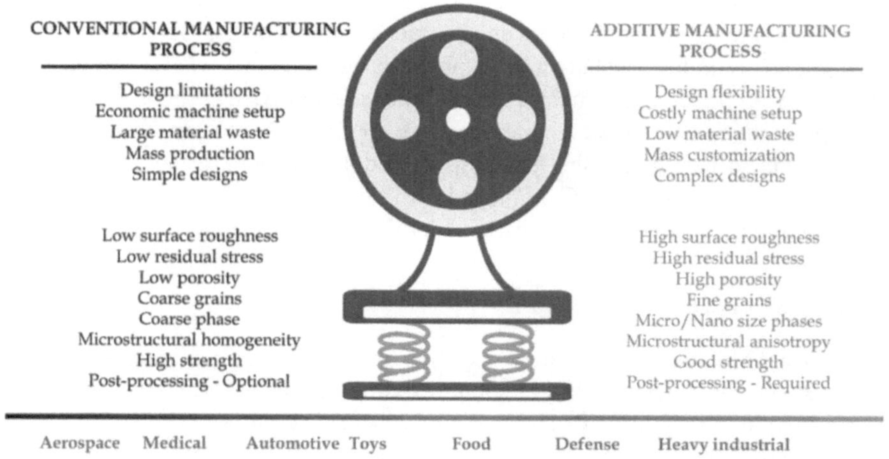

FIGURE 4.2 Comparing additively manufactured components with the traditional ones.

4.1.2 AM Process Limitations

Despite handling a wide range of materials and producing unique designs with properties comparable to wrought or superior to cast counterparts, metal additive manufacturing (AM) has yet to expand to accommodate inherent AM process challenges like poor surface roughness, staircase effect, dimensional accuracy, low fatigue strength, low wear and corrosion resistance, high structural porosity, and high residual stresses [8, 11]. Even when using various AM machines, it might be challenging to achieve identical results because surface anomalies can vary greatly from one another. These variations were seen even when the same 3D printer was used with various feedstock materials. As a result, there haven't been many instances of regulating surface roughness and mechanical properties simultaneously [7, 8]. These anomalies are primarily the result of a poor understanding of process dynamics resulting from complex metallurgical and thermos-physical phenomena.

It is crucial to assess the strong bonding forces in processing zones and the quick solidification phenomenon under an extremely high temperature gradient in the SLM process. Further research is also needed on the changes in thermal stress under cyclic settings and the evolution of the internal structure of the pieces. Internal flaws including balling, porosity, fractures, powder aggregation, and thermal stress would manifest themselves between various printing layers during the manufacturing operations. The interior microstructure and mechanics of the finished products are significantly impacted by these flaws [3, 12–15]. This ultimately restricts the uses of AM materials to simple formats unless final components are placed through extra production steps, such as thermal, chemical, or mechanical-based post-processing activities.

A significant improvement in the overall mechanical and microstructural characteristics of AM products has already been seen after thermal processing. Heat treatments, however, are inadequate to deal with problems brought on by the uneven surface texture, which is frequently associated with poor corrosion and wear performance. Therefore, it is crucial to subject the surfaces of AM component surfaces to mechanical processing to enable surface layer strengthening. Many research groups have worked on various post-processing solutions to alleviate these constraints [7].

To impart desirable mechanical and physical qualities into PBF components, a variety of post-processing techniques are available. It has been found that the mechanical properties of the treated component are significantly influenced by the final nature of the surface layer. Shot-peening, VSF, drag finishing, grinding, polishing, electrochemical polishing, CNC milling, abrasive flow machining, electroplating, and micromachining are some of the most popular finishing techniques on the market.

4.1.3 AM Defects

To produce AM structures that are defect-free and mechanically sturdy, it is crucial to have a thorough understanding of the processing parameters, feedstock materials, and a variety of common faults and how to fix them. The purpose of this is to allow them to modify the mechanical and microstructural performances as a whole. The following section briefly discusses the most common AM defects.

4.1.3.1 Surface Roughness

Surface roughness in AM components is dependent on a variety of interrelated factors, including material feedstock, part design, process type, and process parameters [7]. High surface roughness is typically caused by inadequate melting of the powder, balling owing to Raleigh instability, layer delamination, the stair-step effect during scanning of curved or inclined surfaces, or melt tracks [7, 9]. Insufficient heat input has been cited in a number of studies as the main cause of PBF component surfaces with high surface roughness [16]. There is a limit to how effectively complete powder melting, which is accomplished at high heat input, can minimize surface roughness because extremely high heat can also degrade surface quality due to high thermal stresses and non-uniform solidification rate. Surface imperfections can be greatly reduced by using post-processing techniques, although at an additional expense [7, 9].

4.1.3.2 Porosity

Porosity is a common process-induced AM defect that can be seen in even the best-optimized AM structures [7, 17]. They could have a spherical or ad hoc shape, be disjointed or interconnected, or appear open or closed [18–21]. The most frequent causes of SLM-induced porosity include absence of fusion pores brought on by insufficient powder melting, layer bonding, gas entrapment, melt-pool instability, or key-hole pores led by excessive energy density [18, 20, 22, 23]. Large, asymmetric lack-of-fusion pores are the most dangerous of these pores, leading to inferior mechanical and electrochemical performances. Hot isostatic pressing (HIP) and other post-processing techniques can close internal pores, especially tiny spherical ones, however they are not appropriate for removing surface porosity. HIP is also expensive and increases the time and expense of producing a part [24].

4.1.3.3 Residual Stress and Distortion

Locked-in stresses are the residual stresses in AM structures. The concentrated energy density distribution is a product of the PBF process's distinctive manufacturing setup [8]. Due to the disparity in volume between the newly created and initial metallurgical phases, the phase transformation during cooling can also lead to residual stresses in AM alloys [25].

High tensile residual stresses might reduce the threshold stress necessary for fracture initiation and development when paired with applied stress, leading to fatigue in AM components [26, 27]. Tensile residual stresses in AM components can be reduced using a number of inherent and thermal-based techniques, including preheating the substrate, reducing deposition length or scanning islands, spiraling-in scanning strategy, accelerating scanning, lowering layer height, or applying post-heat treatments [28]. The additional methods to lower residual tensile stresses from AM components include mechanical post-processing procedures including shot-peening, drag-finishing, and finished machining [7, 29, 30].

4.1.3.4 Microstructure

The additively made materials are heterogeneous in comparison to the homogeneous microstructure of the conventional components. The segmented melt pool morphology and non-uniform solidification rate of the AM process preclude the alloying components from dispersing uniformly throughout the surface [31]. A non-uniform distribution of elements occurs in and around the melt pool as a result of the alloying elements segregating at the grain boundaries at rapid cooling rates [32]. Additionally, the directed grain development pattern in the direction of the highest heat gradient in combination with epitaxial grain growth during the solidification of the melt pool creates the textured microstructure. Metal texture is a severe problem since it affects the overall mechanical and electrochemical performances of AM components [26]. If applied, the additional post-treatments to homogenize the microstructure and elemental distribution raise the cost of making AM materials, which is counterproductive to the notion of cost-efficiency.

4.2 POST-PROCESSING OF AM PARTS

Support removal is the first post-processing step in any metal-based AM process. After part fabrication, metal or ceramic structures are left inside a chamber surrounded by unsintered powder particles to minimize part distortion. Later, they are mechanically removed using milling, band-saws, cut-off blades, wire-EDM, and other metal cutting techniques [9]. Process traces from AM procedures frequently appear as stair steps, powder adhesion, fill patterns, and after support removal. To some extent, stair stepping can be reduced by using thin layers at the penalty of longer construction times. Part orientation, powder morphological changes, and scan patterns can be used to regulate powder adhesion [7, 33, 34]. Depending on the need, intended use, and kind of material, sandblasting, sanding, grinding, hand-polishing, coating, or polishing may be used to create matte or lustrous finishes to AM surfaces. More advanced post-processing methods, including as shot-peening, abrasive flow machining, finished machining, drag finishing, vibratory surface finishing, ultrasonic shot-peening, electrochemical finishing, and others, can be used to achieve even better surface finishes and texture enhancements [6, 8]. Following sections will discuss in detail the available thermal and non-thermal post-processing technique to change and modify additive materials surfaces and bulk properties.

4.3 THERMAL POST-PROCESSING

Metal producers frequently use heat treatments to reduce residual stresses, crack formation, and for homogenizing microstructure. Using methods like solution heat treatment (SHT), hot isostatic pressing (HIP), and T6 heat treatment, for instance, one can greatly enhance the quality of the finished product [10, 35]. The effects of heat post-processing on the mechanical, microstructural, and electrochemical performances of AM components have been thoroughly examined by many researchers [36–38]. While barely affecting surface features, heat applications primarily influence the bulk properties of AM components. This indicates that the heat-based treatments are unlikely to address the surface roughness of AM components. Even while some study indicates that surface roughness indices alter following heat treatment, it is difficult to quantify them at first, and variations were also moderate [7, 8].

Previous publications have clearly specified the presence of pores and cracks in AM materials [9]. Regulating LPBF process parameters can significantly improve part density, but as of now, there is no way to totally remove porosity from AM materials [35]. Compared to conventional equivalents, porosity is typically high in AM materials, and this is attributed to unique AM production methodology and high reactivity of metal powders. Pores are often seen near the grain or melt-pool boundaries either due to the gas entrapment or incomplete powder melting. Given the high metal powder reactivity to surrounding air, measures are taken to contain the oxide limit inside the process chamber. Metal oxidation can be significantly reduced by maintaining inert gas pressure (oxygen, nitrogen, etc.) or processing PBF in a vacuum [39]. Even with the utmost precautions, some amounts of oxide remains inside the chamber [40]. As they stay entrapped during the atomization process or enter during transportation, oxide particles inside a powder bed are challenging to remove. Metal powder readily interacts with accessible oxides when exposed to high temperatures, creating metal oxides that either escape during the solidification process or remain trapped, leading to unwanted pores or cracks after solidification [1, 8, 41].

Except for HIP, conventional heat treatments are unfit to resolve issues pertaining to part density. By combining high temperature and pressure to densify finished components, HIP is a widely used thermomechanical treatment method. In a closed chamber containing a high-pressure inert gas, it normally warms up to a temperature of between 1000 and 2000 °C at a working pressure of 200 MPa. Components in HIP are consistently under pressure from all directions to improve the overall density. Quick production cycles, low energy use, and optimal resource usage characterize HIP. The intrinsic flaws and porosity of the PBF pieces can be rectified or eliminated using the HIP process [6, 42, 43].

Investigations have shown that Ti-6Al-4V components manufactured with EBM after HIP have a noticeably higher fatigue strength [42, 44, 45]. It is the outcome of the microstructure's coarsening and the interior pores' closure during the high temperature treatment. In another case, [46] observed a reduced stress corrosion crack (SCC) growth rate in SLM 316L post HIP. However, [47] found that HIP had a detrimental effect on SLM 316L's electrochemical performance in a 0.5M H2SO4 solution because of the rise in corrosion nucleation sites brought on by oxide coalescence in the pores or by lingering fractures from HIP. When compared to as-built materials, HIP creates structures with an extremely high ductility of about 8% [43]. According to [48], adding HIP to other common heat treatment procedures can further reduce pore volume by orders of magnitude [47]. The authors of [49] observed improvements in the IN718 alloy's high strength, hardness, and 15–19% ductility after HIP + conventional heat treatment. While pores on the surface are insensitive to the HIP impact, those present in the bulk are obviously filled in by creep and diffusion processes, as noticed by [50].

Compared to HIP, the influence of conventional heat post-processing on the mechanical and electrochemical behavior of AM materials has so far mostly been assessed [7, 9]. According to [1, 37, 51, 52], aging or solution temperatures can affect the general mechanical and microstructural characteristics. In the example of LPBF IN718, solution heat treatment (SHT) was found to remove the undesirable laves phase by dissolving it into a precipitate strengthening phase, resulting in a hardness increase from 365 to 470 Hv [53–55]. High hardness is boosted by the emergence of intermetallic compounds after aging at the grain boundaries, such as Mg_2Si in AlSi10Mg alloy, Ni-rich intermetallics in maraging steel, and γ' and γ'' phases in IN718 [56–60]. Both grain growth and the spread of intergranular cracks can be restricted by intermetallic particles. Similar results were obtained in the as-built LPBF maraging steel structures, whose strength was seen to increase by almost two-fold following the heat treatment (solution + aging). This results from the precipitation of hard intermetallic compounds at the aging temperature of about 500 °C [15].

The hardness and strength of AM materials though were found dramatically increasing after the aging treatment, the hard and brittle intermetallics lead to poor ductility in them. Low-temperature aging has very little microstructural impact on grain or melt-pool morphology, which continues to display the as-built AM characteristics [61]. However, solution temperature has the ability to totally homogenize a heterogeneous as-built microstructure with coarse grains devoid of essential intermetallic particles. Although the homogenized microstructure at solution temperature increases the material's ductility, the strength significantly decreases and is comparatively high in non-ferrous alloys like AlSi10Mg and Ti-6Al-4V. Therefore, post-aging of solutionized materials is commonly performed in order to boost their strength through intermetallic precipitation. This leads to a balance between mechanical hardness and enhanced ductility, which improves the mechanical performance of AM materials as a whole [49, 57, 62, 63].

When handling metal alloys, heat treatment temperature regulation is crucial because it has a significant impact on the overall hardness of AM materials. Alloy hardness may be impacted by prolonged exposure to high temperatures for extended periods of time. This is well-documented in the instance of Inconel, where poor hardness was reported as a result of the dissolution on strengthening phases like γ' and γ'' [64]. Similar conclusions can be reached in the instance of maraging steel, where lower hardness was attributed to high temperature exposure close to the austenite range. When heated for a long time or close to the austenite temperature range, Ni, a key alloying element in maraging steel, is said to dissolve into the Fe matrix [65]. Temperature treatment and its duration can be used to regulate the phases' size and number [66]

The effect of heat treatment on hardness was seen to diminish in the case of lighter AM alloys such AlSi10Mg, 316L, and Ti-6Al-4V alloys due to the grain coarsening. However, when ductility was taken into account, the total mechanical effect was shown to be substantially better. It occurs as a result of the high temperature precipitation of more ductile phases, as in the instance of the Ti-6Al-4V alloy, where the dominant β phase was discovered to be precipitating from a brittle α' martensite. Comparing the results to the as-built, as-cast, or as-wrought specimens, the total strength and improvement in ductility are much greater [19, 67].

The mechanical properties of AM components are frequently improved through heat treatment in order to make them suitable for energy-efficient applications. For instance, [68] ascribed stable and compact passive film development in SLM 316L SS at 950 °C to dislocation elimination and homogeneous alloying element distribution.

To increase mechanical strength and decrease susceptibility to corrosion, it is crucial to mitigate tensile residual stress, and [69] confirmed this for compressive residual stress.

4.4 NON-THERMAL POST-PROCESSING

LPBF materials have not received as much attention from non-thermal post-processing operations as conventional materials have. Only recently PBF material testing utilizing non-thermal techniques has increased dramatically as a result of recent advancements in additive material handling. It is because the LPBF materials are becoming more widely used. Moreover, their physical and structural development is crucial to their successful final application.

An exhaustive examination of the literature on surface roughness has conclusively shown that the PBF process optimization and thermal heat treatments are not suitable for producing smooth surfaces like those on wrought materials. Therefore, it's crucial to investigate non-thermal post-processing techniques to get the appropriate outcome needed for applications that need a high level of surface finish. Non-thermal techniques that interact with materials to create surfaces with improved strength and finishes can be mechanical, electrical, or chemical in nature. Due to the physical contact of non-thermal media with the component surface, non-thermal procedures and material interaction are typically restricted to the surface or subsurface regions. Depending on the process, alloy material, and structural topography, the material media interaction is generally confined to material erosion and compression. Due to varying process methodology and approaches to material interaction, not all procedures yield the same results. Therefore, it is crucial to carefully consider the entire cost, time, and production steps when choosing materials and post-processing techniques to get the desired output [1, 70, 71].

The strength and surface polish of AM components have frequently been tested using non-thermal post-processing techniques. Nearly all of these techniques are successful in reducing surface roughness, but not all of them are sufficient to provide added strength or hardness to AM materials. For instance, ECP, polishing, and grinding can yield excellent surface finishes, but compared to the outcomes frequently attained through operations like shot-peening, FM, DF, and VSF, increases in surface hardness are negligible. By bringing peaks closer to the actual surface, non-thermal procedures can effectively reduce surface asperities. Depending on the methods, material, and shape of the LPBF component, the peaks are either melted (electrochemical or laser polishing), eliminated (polishing, grinding, FM), or compressed (shot-peening, DF, or VSF). These post-processing methods can be employed individually or collectively to produce desired effects. For instance, the impact on the final surface quality is significantly stronger when ECP and FM are combined than when ECP is used alone on AM materials [72]. Additionally, it was discovered that AlSi10Mg alloy surface roughness significantly decreased when shot-peening was combined with mechanical polishing or the ECP process [73].

For surface roughness, not just the non-thermal process type but its process optimization is equally essential to induce desired results. Non-thermal process

TABLE 4.1
Typical Process Parameters for Different Non-Thermal Processes

Non-Thermal Process Type	Process	Processing Parameters
Mechanical based	Vibratory surface finishing	Granules type, size and shape, lubrications, process duration, vibration frequency
	Drag finishing	Rotational speed, media surface, media pressure, media velocity, lubricants
	Shot-peening (SP)	Air pressure, mass flow, impact angle, exposure time
	Centrifugal disc finishing (CDF)	Location of the workpiece, oscillation, rotational movements
	Some other process: laser polishing (LP), milling (M) ultrasonic cavitation abrasive finishing, grinding (Gr), polishing (P) hydrodynamic, rolling, bead blasting, ultrasonic nano-crystal surface modification,	
Electro-chemical based	Electro-chemical Polishing (ECP)	voltage, electrode gap, electrolyte temperature
	Electroplating	Microjet fly-height, electrolyte flow rate, applied current
	Chemical etching	etching agent, time, temperature
	Some other process: coating, chemical polishing, hybrid treatments	

optimization can be achieved by regulating size, shape, concentration, and distribution of interacting media and process configuration. Depending on the process, these parameters can be categorized. Table 4.1 shows the typical process parameters for different non-thermal processes.

After applying post-processing, experimental studies have demonstrated a striking improvement in the surface roughness and mechanical performance of additive materials. For instance, regardless of the ball material, such as ceramic, steel, or glass beads, the shot-peening process reduces the surface roughness and enhances the microhardness of the as-built AlSi10Mg surface. Given that the parameters of the shot-peening process might vary depending on the material being processed, harder materials require more energy to achieve the same results [74]. The finished machining process can produce surface finish similar to wrought material in a short span. The authors of [1] observed nearly similar surface roughness development on using two different FM parameters, which was close to the as-received wrought material. Drag finishing (DF) and vibratory surface finishing (VSF) processes were also found to significantly improve the surface roughness and mechanical performance of 316L. The FM, SP, LP, or ECP processes can accomplish similar outcomes in a short amount of time as opposed to the DF or VSF processes, which can take several hours to treat the surface [75, 76].

The duration is also an important post-processing parameter. With no further modification to the roughness values, prolonged use of the shot-peening technique might cause cracks in the surface [77, 78]. Furthermore, prolonged usage in some operations can seriously harm the part. For instance, excessive material loss from finished machining or material softening after electrochemical polishing are both possible outcomes [79]. Electrochemical polishing (ECP) softens the microsurface by releasing tensions and removing excess material. When attempting to achieve a smooth surface texture, this is very helpful, but it's prolonged use is harmful to the surface hardness [79, 80].

The challenge that post-processing approaches are intended to address is that the majority of these procedures (SP, FM) were only found to be compatible with the standard planar geometry, leaving the more intricate AM profile unchanged. To overcome these limitations and increase the surface roughness of geometrically challenging, high-value AM products, mass finishing methods (DF, VSF) are gaining popularity. Even though the advantages of mass finishing techniques have been the subject of numerous studies, much more analysis is still required to standardize their impact on AM products. However, nearly all mechanical-based post-processing techniques are inadequate when taking into account complex structures like internal cooling channels or complicated lattice materials. This issue is somewhat addressed by ECP methods, but for a successful outcome, treatment time must be customized. Prolonged exposure to chemical solutions can cause material surfaces to become abnormally soft, which can result in fracture development and poor mechanical and electrochemical performances. The LPBF processing settings and the post-processing parameters must both be optimized in order to get a high level of surface quality. The low mechanical strength is caused by the significant surface abrasion and poor densification. High surface roughness has a negative impact on wear and corrosion resistance as well.

As was already established, one of the main microstructural flaws in AM materials is porosity. Areas around pores are frequently susceptible to high stresses, making their removal quite simple. If they are not kept under control, they might serve as a breeding ground for unfavorable ion concentration, which might lead to the development of galvanic cells. As a result, the material might experience excessive material damage, and cracks might have a chance to deepen. Additionally, under stress, crack propagation through pores is made simple, which has a negative impact on the material's mechanical strength and fatigue resistance.

Moreover, the surface becomes strain-hardened as a result of mechanical processes like shot-peening, DF, VSF, and others. This causes the grains on the surface and within the subsurface region to get refined due to their plastic deformation by the high pressure media particles. As a result, the microhardness of the surface increases compared to the non-treated surface. The Hall-Petch equation states that the surface strength is inversely proportional to the grain size [81]. For instance, 316L surfaces following FM, DF, and VSF demonstrated finer grain morphology, leading to higher microhardness near the surface than the bulk. The effect of FM being most noticeable, about 20 to 30% higher than the as-built surface. Following FM operation on a maraging steel alloy, [1] showed a considerable improvement in surface hardness from 8–10 to 0.5 μm, approximately. By milling a DED ASTM A131 steel part at a high cutting speed, [82] found that the surface roughness was reduced from 22.78 to 0.6 μm. However, unlike [76], they found no change in mechanical hardness. Following the magnetic abrasive finishing (MAF) procedure, [83] found a 75.7% improvement in the elimination of balling and unmelted borders in 316L SS. The authors of [84] discovered a surface roughness reduction from 5.02 to 2.93 m using ultrasonic abrasive polishing. The surface roughness of AlSi10Mg components is similarly reduced by grinding and MAF, going from 7 to 0.155 m [85]. When abrasive flow machining was employed to treat the AM surface, [86] observes a 26% improvement in fatigue strength as a result of the decrease in residual stresses. These procedures may normally be used to treat hard AM surfaces, but careful adjustment of the procedure's parameters is needed to treat lattice struts and interior areas without damaging the delicate lattice columns [87, 88].

After the DF treatment of the maraging steel alloy, [32] showed an improvement in surface hardness. The effects of grinding, electrolytic etching, sandblasting, and plasma polishing on LPBF 316L SS were examined by [89]. They discovered that grinding had the best results when the surface roughness (Ra) was reduced from 15 to 0.34 µm. Similar to this, multiple researchers showed that applying the shot-peening approach to various materials increased the surface hardness and refined the grain appearance [78, 90–92]. References [76, 93, 94] conducted a number of post-processing studies on additive samples and found that AM samples performed better in terms of mechanical strength and wear resistance than their as-built equivalents. Following the bulk finishing process, [32, 95] found that LPBF samples had increased mechanical performance. The authors of [96] found that aluminum components after VSF and shot-peening had similar fatigue performance. Furthermore, in terms of surface quality, VSF samples fared better than shot-peened samples. Other researchers [97] have also corroborated these findings. Due to its ability to handle multiple components at once and the fact that many of its variants don't have complex tool path designs or clamping fixtures, mass finishing is suitable for large and medium-sized batches [98].

When a surface deforms plastically, essential compressive stresses are created that act as a buffer against the applied load [6, 7]. As refined grains considerably limit dislocation movement compared to unrefined grains, this automatically increases surface strength against crack nucleation or propagation, giving the structure high mechanical and fatigue strength [92]. Even if a crack does start on the surface, it will spread much more slowly because to the presence of fine-grain structures and lower tensile strains [99].

When surfaces are polished with a laser, the surrounding pores and valleys are reduced as the laser remelts the surface peaks to level them without changing the bulk properties [100]. As of now, different AM materials can be polished using laser polishing technology. This was clearly demonstrated when Ti-6Al-4V and 316L alloys were polished using a laser, with 80% and 60% of the surface being smoothed upto a depth of 100 µm, respectively [101, 102]. Similarly, a roughness reduction from 195 to 75 nm was observed in 304 SS after laser polishing [103]. Effect of laser polishing was also positively observed on hardness in a commercial alloy (LaserForm ST-100©) without cracks or HAZ regions [102]. Research is still ongoing, however a laser polishing process optimization can further anticipate better outcomes for many other additive materials [104, 105]. Table 4.2 presents the comparison of the properties of the different materials before and after post processing.

TABLE 4.2
Comparison of the Material Properties Before and After Post-Process

	Surface Roughness (µm)		Heat Treatment (Yield Strength, MPa)		Heat Treatment (Hardness)	
Material	As-Built	After Post-Process	As-Built	After Post-Process	As-Built	After Post-Process
AlSi10Mg	6-12	SP:2–6, Machined: ~0.22 [78, 91, 106]	231–268	100–300 [107–109]	95–125	60–90 [107, 110]
Maraging steel	10-20	SP: 4–6.5 [111]	10–1200	2100–2450 [112, 113]	300–400	520–680 [40, 114]
316L	7-24	ECP:9.3, LP: ~2–8, Gr:0.05 [106, 115]	480–760	550–600 [63, 116]	220–280	210–275 [117]
Ti6Al4V	6-18	M+P:0.1–0.3, [118]	1150–1267	950–1050 [119, 120]	420–613	350–410 [19, 121]
CoCr-Mo	5-15	P:0.01, Gr: ~4, LP: ~2 [122, 123]	591–655	400–920 [124]	395–420	280–470 [125]

4.5 COMBINATION OF THERMAL AND MECHANICAL METHODS

The effect of the combination of thermal and mechanical properties on LPBF materials has been studied in the past. On DED prepared Inconel 718 super-alloy, [126] studied different post-processing techniques to improve its surface integrity. Here, first machining and then heat treatment or first heat treatment and then machining were chosen as two approaches to post-treat the Inconel alloy. It was shown that only machining produced specimens with surfaces that were low in surface roughness and high in surface hardness, with surface shrinkage predominating. In contrast, machining followed by heat treatment led to poor machinability because of a harder surface brought on by the precipitation of intermetallic particles. However, performing machining first and then the heat treatment method revealed a notable increase in machinability and a significant hardness rise [126].

The impact of heat treatment and drag finishing on the surface integrity and corrosion behavior of LPBF maraging steel specimens was assessed in a different investigation. MS1 material's hardness was enhanced by heat treatment by a factor of roughly 2, while drag finishing lessened surface roughness. When compared to the structure as it was manufactured, the effects of drag-finishing and heat treatment improved the structure's overall physical, mechanical, and electrochemical qualities.

4.6 CORROSION AND WEAR

A detailed discussion on the effect of post-processing applications on LPBF materials have being discussed in [6–8]. This present section will briefly present the overview of the effect of post-processing applications on different LPBF materials.

Heat treatment of AM metallic parts has significant effects on corrosion. The heat treatments usually applied to LPBF AlSi10Mg alloys are stress relieving (SR) and T6 heat treatments. The SR heat treatment does not affect the microstructure much, but the T6 heat treatment has a dramatic effect on the AM microstructure. After T6 heat treatment, the as-built microstructure and melt-pool morphology are completely transformed into a course microstructure with coarse Si particles. This greatly affects the corrosion kinetics of AlSi10Mg surfaces. Given a significant potential difference between the silicon and the aluminum matrix, coarse Si particles can form several microgalvanic cells that causes in an aggressive surface erosion post T6 treatment [127].

The heat treatments applied for LPBF MS1 are aged and solution (SHT) heat treatments. Similar to aluminum, aging only facilitates in precipitating hard intermetallic particles without causing a significant change in the overall surface microstructure. On applying SHT heat treatment, the melt pool microstructure disappears and a lath-like martensite structure is formed. Studies have shown that the as-built microstructure shows the highest corrosion resistance, while SHT treatment reduces the LPBF MS1 corrosion resistance. The probable reason for this is that the martensite form formed after SHT reduces the corrosion resistance [128]. A similar effect was also noticed by [32], as discussed above.

When it comes to 316L material, its microstructure is completely austenitic, which remain invariant to heat treatments as high as 800 °C. Temperatures higher than 800 °C can cause surface recrystallization, which can result in a significant drop in the pitting potential of 316L material [129, 130]. After heat treatment, LPBF Ti6Al4V microstructure was found changing from the metastable α' phase turns into $\alpha + \beta$ structure. [131] reported a positive impact of heat treatment on the corrosion resistance of LPBF titanium alloy. This is explained by the formation of a protective passive film formed on the surface.

4.7 CONCLUSION

AM technologies have gained importance in recent years. It will come as no surprise that we will increasingly see AM technologies in the manufacturing sector in the future. The performance of AM materials is highly dependent on the post-processing applied to the parts. For the control of surface roughness, porosity, residual stresses and microstructure, a number of post-treatments must be applied to the parts produced with AM. For this, post-treatments such as heat treatments, HIP, non-thermal treatments, drag finishing, shot-peening, polishing are applied to metallic AM parts. In the future, developments in all these post-treatment technologies will be important to meet the growing potential in the AM industry.

REFERENCES

1. Tascioglu, E., Khan, H. M., Kaynak, Y., Coşkun, M., Tarakci, G., & Koç, E. (2021). Effect of aging and finish machining on the surface integrity of selective laser melted maraging steel. *Rapid Prototyping Journal*, *27*(10), 1900–1909. https://doi.org/10.1108/RPJ-11-2020-0269

2. Khan, H. M., Çalışkan, C. İ., & Bulduk, M. E. (2022). The novel hybrid lattice structure approach fabricated by laser powder bed fusion and mechanical properties comparison. *3D Printing and Additive Manufacturing*. https://doi.org/10.1089/3dp.2022.0224
3. Khan, H. M., Waqar, S., & Koç, E. (2022). Evolution of temperature and residual stress behavior in selective laser melting of 316L stainless steel across a cooling channel. *Rapid Prototyping Journal*, *28*(7), 1272–1283. https://doi.org/10.1108/RPJ-09-2021-0237
4. Çalışkan, C. İ., Khan, H. M., Özer, G., Waqar, S., & Tütük, İ. (2022). The effect of conformal cooling channels on welding process in parts produced by additive manufacturing, laser powder bed fusion. *Journal of Manufacturing Processes*, *83*, 705–716. https://doi.org/10.1016/j.jmapro.2022.09.036
5. Khan, H. M., Dirikolu, M. H., & Koç, E. (2018). Parameters optimization for horizontally built circular profiles: Numerical and experimental investigation. *Optik*, *174*(August), 521–529. https://doi.org/10.1016/j.ijleo.2018.08.095
6. Majeed, M., Khan, H. M., Wheatley, G., & Situ, R. (2022). Influence of post-processing on additively manufactured lattice structures. *Journal of the Brazilian Society of Mechanical Sciences and Engineering*, *44*(9), 389. https://doi.org/10.1007/s40430-022-03703-8
7. Khan, H. M., Karabulut, Y., Kitay, O., Kaynak, Y., & Jawahir, I. S. (2021). Influence of the post-processing operations on surface integrity of metal components produced by laser powder bed fusion additive manufacturing: a review. *Machining Science and Technology*, *25*(1), 118–176. https://doi.org/10.1080/10910344.2020.1855649
8. Khan, H. M., Özer, G., Yilmaz, M. S., & Koc, E. (2022). Corrosion of additively manufactured metallic components: A Review. *Arabian Journal for Science and Engineering*, *47*(1), 5465–5490. https://doi.org/10.1007/s13369-021-06481-y
9. DebRoy, T., Wei, H. L., Zuback, J. S., Mukherjee, T., Elmer, J. W., Milewski, J. O., … Zhang, W. (2018). Additive manufacturing of metallic components–Process, structure and properties. *Progress in Materials Science*, *92*, 112–224. https://doi.org/10.1016/j.pmatsci.2017.10.001
10. Zadpoor, A. A. (2019). Mechanical performance of additively manufactured meta-biomaterials. *Acta Biomaterialia*, *85*, 41–59. https://doi.org/10.1016/j.actbio.2018.12.038
11. Khan, H. M., Yilmaz, M. S., Karabeyoğlu, S. S., Kisasoz, A., & Özer, G. (2022). Dry sliding wear behavior of 316L stainless steel produced by laser powder bed fusion: A comparative study on test temperature. *Materials Today Communications*, 105155. / https://doi.org/10.1016/j.mtcomm.2022.105155
12. Wang, C. G., Zhu, J. X., Wang, G. W., Qin, Y., Sun, M. Y., Yang, J. L., … Huang, S. K. (2022). Effect of building orientation and heat treatment on the anisotropic tensile properties of AlSi10Mg fabricated by selective laser melting. *Journal of Alloys and Compounds*, *895*, 162665. https://doi.org/10.1016/j.jallcom.2021.162665
13. Waqar, S., Sun, Q., Liu, J., Guo, K., & Sun, J. (2021). Numerical investigation of thermal behavior and melt pool morphology in multi-track multi-layer selective laser melting of the 316L steel. *The International Journal of Advanced Manufacturing Technology*, *112*(3), 879–895.
14. Waqar, S., Guo, K., & Sun, J. (2021). FEM analysis of thermal and residual stress profile in selective laser melting of 316L stainless steel. *Journal of Manufacturing Processes*, *66*, 81–100. https://doi.org/10.1016/j.jmapro.2021.03.040
15. Özer, G., Khan, H. M., Tarakçi, G., Yilmaz, M. S., Yaman, P., Karabeyoğlu, S. S., & Kisasöz, A. (2022). Effect of heat treatments on the microstructure and wear behaviour of a selective laser melted maraging steel. *Proceedings of the Institution of Mechanical*

Engineers, Part E: Journal of Process Mechanical Engineering, 09544089221093994. https://doi.org/10.1177/09544089221093994

16. Everton, S. K., Hirsch, M., Stravroulakis, P., Leach, R. K., & Clare, A. T. (2016). Review of in-situ process monitoring and in-situ metrology for metal additive manufacturing. *Materials and Design*, *95*, 431–445. https://doi.org/10.1016/j.matdes.2016.01.099
17. Keles, O., Shuja, S. Z., Yilbas, B. S., Al-Qahtani, H., Hassan, G., Adesina, A. Y., … Al-Sharafi, A. (2014). Additive manufacturing of Ti-alloy: Thermal analysis and assessment of properties. *Advances in Mechanical Engineering*, *12*(6), 1687814020933068. https://doi.org/10.1177/1687814020933068
18. Qiu, C., Wang, Z., Aladawi, A. S., Al Kindi, M., Al Hatmi, I., Chen, H., & Chen, L. (2019). Influence of laser processing strategy and remelting on surface structure and porosity development during selective laser melting of a metallic material. *Metallurgical and Materials Transactions A*, *50*(9), 4423–4434.
19. Vilaro, T., Colin, C., & Bartout, J. D. (2011). As-fabricated and heat-treated microstructures of the Ti-6Al-4V alloy processed by selective laser melting. *Metallurgical and Materials Transactions A: Physical Metallurgy and Materials Science*, *42*(10), 3190–3199. https://doi.org/10.1007/s11661-011-0731-y
20. Khairallah, S. A., Anderson, A. T., Rubenchik, A. M., & King, W. E. (2016). Laser powder-bed fusion additive manufacturing: Physics of complex melt flow and formation mechanisms of pores, spatter, and denudation zones. *Acta Materialia*, *108*(10), 36–45. https://doi.org/10.1201/9781315119106
21. Schaller, R. F., Taylor, J. M., Rodelas, J., & Schindelholz, E. J. (2017). Corrosion properties of powder bed fusion additively manufactured 17-4 PH stainless steel. *Corrosion*, *73*(7), 796–807.
22. Brooks, J. W., Qiu, C., Attallah, M. M., Panwisawas, C., Ward, M., & Basoalto, H. C. (2015). On the role of melt flow into the surface structure and porosity development during selective laser melting. *Acta Materialia*, *96*, 72–79. https://doi.org/10.1016/j.actamat.2015.06.004
23. Qiu, C., Adkins, N. J. E., & Attallah, M. M. (2016). Selective laser melting of Invar 36: microstructure and properties. *Acta Materialia*, *103*, 382–395.
24. Ertuğrul, O., Öter, Z. Ç., Yılmaz, M. S., Şahin, E., Coşkun, M., Tarakçı, G., & Koc, E. (2020). Effect of HIP process and subsequent heat treatment on microstructure and mechanical properties of direct metal laser sintered AlSi10Mg alloy. *Rapid Prototyping Journal*.
25. Shahriari, A., Khaksar, L., Nasiri, A., Hadadzadeh, A., Amirkhiz, B. S., & Mohammadi, M. (2020). Microstructure and corrosion behavior of a novel additively manufactured maraging stainless steel. *Electrochimica Acta*, *339*, 135925. https://doi.org/10.1016/j.electacta.2020.135925
26. Mohtadi-Bonab, M. A. (2019). Effects of different parameters on initiation and propagation of stress corrosion cracks in pipeline steels: A review. *Metals*, *9*(5), 1–18. https://doi.org/10.3390/met9050590
27. Bruycker, E. De, Sistiaga, M. L. M., Thielemans, F., & Vanmeensel, K. (2017). Corrosion Testing of a Heat Treated 316 L Functional Part Produced by Selective Laser Melting. *Materials Sciences and Applications*, *08*(03), 223–233. https://doi.org/10.4236/msa.2017.83015
28. Gao, M., Wang, Z., Li, X., & Zeng, X. (2013). The effect of deposition patterns on the deformation of substrates during direct laser fabrication. *Journal of Engineering Materials and Technology*, *135*(3).

29. Patterson, A. E., Messimer, S. L., & Farrington, P. A. (2017). Overhanging features and the SLM/DMLS residual stresses problem: Review and future research need. *Technologies*, *5*(2), 15. https://doi.org/10.3390/technologies5020015
30. Zhuo, L., Wang, Z., Zhang, H., Yin, E., Wang, Y., Xu, T., & Li, C. (2019). Effect of post-process heat treatment on microstructure and properties of selective laser melted AlSi10Mg alloy. *Materials Letters*. https://doi.org/10.1016/j.matlet.2018.09.109
31. Huang, L., Wang, X., Zhao, X., Wang, C., & Yang, Y. (2021). Analysis on the key role in corrosion behavior of CoCrNiAlTi-based high entropy alloy. *Materials Chemistry and Physics*, *259*, 124007. https://doi.org/https://doi.org/10.1016/j.matchemphys.2020.124007
32. Khan, H. M., Özer, G., Tarakci, G., Coskun, M., Koc, E., & Kaynak, Y. (2021). The impact of aging and drag-finishing on the surface integrity and corrosion behavior of the selective laser melted maraging steel samples. *Materialwissenschaft und Werkstofftechnik*, *52*(1), 60–73. https://doi.org/https://doi.org/10.1002/mawe.202000139
33. Karthik, R., Elangovan, K., Shankar, S., & Girisha, K. G. (2022). An experimental analysis on surface roughness of the selective laser sintered and unsintered Inconel 718 components using vibratory surface finishing process. *Materials Today: Proceedings*. https://doi.org/10.1016/j.matpr.2022.04.448
34. Tarakçı, G., Khan, H. M., Yılmaz, M. S., & Özer, G. (2022). Effect of building orientations and heat treatments on AlSi10Mg alloy fabricated by selective laser melting: microstructure evolution, mechanical properties, fracture mechanism and corrosion behavior. *Rapid Prototyping Journal*, *28*(8), 1609–1621. https://doi.org/10.1108/RPJ-11-2021-0325
35. Olakanmi, E. O., Cochrane, R. F., Dalgarno, K. W., & E.O. Olakanmi, R. F. C. K. W. D. (2015). A review on selective laser sintering/melting (SLS/SLM) of aluminium alloy powders: Processing, microstructure, and properties. *Progress in Materials Science*, *74*, 401–477. https://doi.org/10.1016/j.pmatsci.2015.03.002
36. Haghdadi, N., Laleh, M., Moyle, M., & Primig, S. (2021). Additive manufacturing of steels: a review of achievements and challenges. *Journal of Materials Science*, *56*(1), 64–107. https://doi.org/10.1007/s10853-020-05109-0
37. Zhang, B., Li, Y., & Bai, Q. (2017). Defect Formation Mechanisms in Selective Laser Melting: A Review. *Chinese Journal of Mechanical Engineering (English Edition)*. https://doi.org/10.1007/s10033-017-0121-5
38. Sander, G., Tan, J., Balan, P., Gharbi, O., Feenstra, D. R., Singer, L., … Birbilis, N. (2018). Corrosion of additively manufactured alloys: A review. *Corrosion*, *74*(12), 1318–1350. https://doi.org/10.5006/2926
39. Eleftherios Louvis, P. F. C. J. S., Louvis, E., Fox, P., & Sutcliffe, C. J. (2011). Selective laser melting of aluminium components. *Journal of Materials Processing Technology*, *211*(2), 275–284. https://doi.org/10.1016/j.jmatprotec.2010.09.019
40. Kempen, K., Yasa, E., Thijs, L., Kruth, J.-P. P., & Van Humbeeck, J. (2011). Microstructure and mechanical properties of selective laser melted 18Ni-300 steel. *Physics Procedia*, *12*(PART 1), 255–263. https://doi.org/10.1016/j.phpro.2011.03.033
41. Weingarten, C., Buchbinder, D., Pirch, N., Meiners, W., Wissenbach, K., & Poprawe, R. (2015). Formation and reduction of hydrogen porosity during selective laser melting of AlSi10Mg. *Journal of Materials Processing Technology*, *221*, 112–120. https://doi.org/10.1016/j.jmatprotec.2015.02.013

42. Tillmann, W., Schaak, C., Nellesen, J., Schaper, M., Aydinöz, M. E. u, & Hoyer, K.-P. (2017). Hot isostatic pressing of IN718 components manufactured by selective laser melting. *Additive Manufacturing, 13*, 93–102.
43. Maier, H. J. J., Tröster, T., Richard, H. A. A., Thöne, M., Leuders, S., Riemer, A., … Maier, H. J. J. (2013). On the mechanical behaviour of titanium alloy TiAl6V4 manufactured by selective laser melting: Fatigue resistance and crack growth performance. *International Journal of Fatigue, 48*, 300–307. https://doi.org/10.1016/j.ijfatigue.2012.11.011
44. Nesma T. Aboulkhair, I. M. C. T. I. A. N. M. E., Aboulkhair, N. T., Maskery, I., Tuck, C., Ashcroft, I., & Everitt, N. M. (2016). The microstructure and mechanical properties of selectively laser melted AlSi10Mg: The effect of a conventional T6-like heat treatment. *Materials Science and Engineering: A, 667*, 139–146. https://doi.org/10.1016/j.msea.2016.04.092
45. Rosenthal, I., Tiferet, E., Ganor, M., & Stern, A. (2015). Post-processing of AM-SLM AlSi10Mg specimens: Mechanical properties and fracture behaviour. *The Annals of" Dunarea de Jos" University of Galati. Fascicle XII: Welding Equipment and Technology, 26*, 33.
46. Lou, X., Song, M., Emigh, P. W., Othon, M. A., & Andresen, P. L. (2017). On the stress corrosion crack growth behaviour in high temperature water of 316L stainless steel made by laser powder bed fusion additive manufacturing. *Corrosion Science, 128*(February), 140–153. https://doi.org/10.1016/j.corsci.2017.09.017
47. Geenen, K., Röttger, A., & Theisen, W. (2017). Corrosion behavior of 316L austenitic steel processed by selective laser melting, hot-isostatic pressing, and casting. *Materials and Corrosion, 68*(7), 764–775. https://doi.org/10.1002/maco.201609210
48. Goel, S., Sittiho, A., Charit, I., Klement, U., & Joshi, S. (2019). Effect of post-treatments under hot isostatic pressure on microstructural characteristics of EBM-built Alloy 718. *Additive Manufacturing, 28*, 727–737. https://doi.org/https://doi.org/10.1016/j.addma.2019.06.002
49. Popovich, V. A. A., Borisov, E. V. V, Popovich, A. A. A., Sufiiarov, V. S., Masaylo, D. V. V, & Alzina, L. (2017). Impact of heat treatment on mechanical behaviour of Inconel 718 processed with tailored microstructure by selective laser melting. *Materials & Design, 131*, 12–22. https://doi.org/10.1016/j.matdes.2017.05.065
50. Bailey, P. G., & Schweikert, W. H. (1976). HIP densification of castings. In *Proceedings of the Superalloys: Metallurgy and Manufacture: Proceedings of the Third International Symposium, Seven Springs, PA, USA* (pp. 12–15).
51. Aboulkhair, N. T., Maskery, I., Tuck, C., Ashcroft, I., & Everitt, N. M. (2016). Improving the fatigue behaviour of a selectively laser melted aluminium alloy: Influence of heat treatment and surface quality. *JMADE, 104*, 174–182. https://doi.org/10.1016/j.matdes.2016.05.041
52. Chen, S., Chen, K., Peng, G., Jia, L., & Dong, P. (2012). Effect of heat treatment on strength, exfoliation corrosion and electrochemical behavior of 7085 aluminum alloy. *Materials and Design, 35*, 93–98. https://doi.org/10.1016/j.matdes.2011.09.033
53. Aydinöz, M. E., Brenne, F., Schaper, M., Schaak, C., Tillmann, W., Nellesen, J., & Niendorf, T. (2016). On the microstructural and mechanical properties of post-treated additively manufactured Inconel 718 superalloy under quasi-static and cyclic loading. *Materials Science and Engineering: A, 669*, 246–258. https://doi.org/https://doi.org/10.1016/j.msea.2016.05.089
54. Zhang, D., Niu, W., Cao, X., & Liu, Z. (2015). Effect of standard heat treatment on the microstructure and mechanical properties of selective laser melting manufactured

Inconel 718 superalloy. *Materials Science and Engineering: A*, *644*, 32–40. https://doi.org/10.1016/j.msea.2015.06.021
55. Wang, Z., Guan, K., Gao, M., Li, X., Chen, X., & Zeng, X. (2012). The microstructure and mechanical properties of deposited-IN718 by selective laser melting. *Journal of Alloys and Compounds*, *513*, 518–523. https://doi.org/10.1016/j.jallcom.2011.10.107
56. Casati, R., Lemke, J. N., Tuissi, A., & Vedani, M. (2016). Aging behaviour and mechanical performance of 18-Ni 300 steel processed by selective laser melting. *Metals*, *6*(9). https://doi.org/10.3390/met6090218
57. Chlebus, E., Gruber, K., Kuźnicka, B., Kurzac, J., Kurzynowski, T., Yang, K. V, … Kurzynowski, T. (2015). Effect of heat treatment on the microstructure and mechanical properties of Inconel 718 processed by selective laser melting. *Materials Science and Engineering: A*, *639*, 647–655. https://doi.org/10.1016/j.msea.2015.05.035
58. Fiocchi, J., Tuissi, A., Bassani, P., & Biffi, C. A. (2017). Low temperature annealing dedicated to AlSi10Mg selective laser melting products. *Journal of Alloys and Compounds*, *695*, 3402–3409. https://doi.org/10.1016/j.jallcom.2016.12.019
59. Lam, L. P., Zhang, D. Q., Liu, Z. H., & Chua, C. K. (2015). Phase analysis and microstructure characterisation of AlSi10Mg parts produced by selective laser melting. *Virtual and Physical Prototyping*, *10*(4), 207–215. https://doi.org/10.1080/17452759.2015.1110868
60. Tian, J., Wang, W., Shahzad, M. B., Yan, W., Shan, Y., Jiang, Z., & Yang, K. (2017). A new maraging stainless steel with excellent strength-toughness-corrosion synergy. *Materials*, *10*(11), 1–11. https://doi.org/10.3390/ma10111293
61. Dos Reis, F., & Ganghoffer, J. F. (2012). Construction of micropolar continua from the asymptotic homogenization of beam lattices. *Computers and Structures*, *112–113*, 354–363. https://doi.org/10.1016/j.compstruc.2012.08.006
62. Niendorf, T., Richard, H. A., Leuders, S., Riemer, A., Tröster, T., & Thöne, M. (2014). On the fatigue crack growth behavior in 316L stainless steel manufactured by selective laser melting. *Engineering Fracture Mechanics*, *120*, 15–25. https://doi.org/10.1016/j.engfracmech.2014.03.008
63. Sistiaga, M. L. M., Nardone, S., Hautfenne, C., & Humbeeck, J. Van. (2016). Effect of Heat Treatment Of 316L Stainless Steel Produced by Selective Laser Melting (SLM), 558–565.
64. Tucho, W. M., Cuvillier, P., Sjolyst-Kverneland, A., & Hansen, V. (2017). Microstructure and hardness studies of Inconel 718 manufactured by selective laser melting before and after solution heat treatment. *Materials Science and Engineering: A*, *689*, 220–232. https://doi.org/10.1016/j.msea.2017.02.062
65. Khan, H. M., Özer, G., Yilmaz, M. S., Tarakçı, G., Mahmood Khan, H., Özer, G., … Tarakci, G. G. (2022). Improvement of corrosion resistance of maraging steel manufactured by selective laser melting through intercritical heat treatment. *Corrosion*, *78*(3), 239–248. https://doi.org/10.5006/3972
66. Murr, L. E., Martinez, E., Amato, K. N., Gaytan, S. M., Hernandez, J., Ramirez, D. A., … Lawrence E. Murr, E. M. K. N. A. S. M. G. J. H. D. A. R. P. W. S. F. M. R. B. W. (2012). Fabrication of metal and alloy components by additive manufacturing: Examples of 3D materials science. *Journal of Materials Research and Technology*, *1*(1), 42–54. https://doi.org/10.1016/S2238-7854(12)70009-1
67. Vrancken, B., Thijs, L., Kruth, J.-P., & Van Humbeeck, J. (2012). Heat treatment of Ti6Al4V produced by selective laser melting: Microstructure and mechanical properties. *Journal of Alloys and Compounds*, *541*, 177–185.

68. Zhou, C., Wang, J., Hu, S., Tao, H., Fang, B., Li, L., … Zhang, L. (2020). Enhanced corrosion resistance of additively manufactured 316L stainless steel after heat treatment. *Journal of The Electrochemical Society*, *167*(14), 141504.
69. Karthik, D., & Swaroop, S. (2017). Effect of laser peening on electrochemical properties of titanium stabilized 321 steel. *Materials Chemistry and Physics*, *193*, 147–155. https://doi.org/https://doi.org/10.1016/j.matchemphys.2017.02.022
70. Khorasani, A. M., Gibson, I., Chegini, N. G., Goldberg, M., Ghasemi, A. H., & Littlefair, G. (2016). An improved static model for tool deflection in machining of Ti–6Al–4V acetabular shell produced by selective laser melting. *Measurement*, *92*, 534–544.
71. Kaynak, Y., Tascioglu, E., Poyraz, Ö., Orhangül, A., & Ören, S. (2020). the effect of finish-milling operation on surface quality and wear resistance of Inconel 625 produced by selective laser melting additive manufacturing. *Advanced Surface Enhancement*. Advanced Surface Enhancement. https://doi.org/10.1007/978-981-15-0054-1_27
72. Tian, C., Li, X., Liu, Z., Zhi, G., Guo, G., Wang, L., & Rong, Y. (2018). Study on grindability of Inconel 718 superalloy fabricated by selective laser melting (SLM). *Strojniski Vestnik/Journal of Mechanical Engineering*, *64*(5), 319–328. https://doi.org/10.5545/sv-jme.2017.4864
73. Uzan, N. E., Shneck, R., Yeheskel, O., Frage, N., Elad, N., Shneck, R., … Frage, N. (2017). Fatigue of AlSi10Mg specimens fabricated by additive manufacturing selective laser melting (AM-SLM). *Materials Science and Engineering A*, *704*(August), 229–237. https://doi.org/10.1016/j.msea.2017.08.027
74. Fortunato, A., Lulaj, A., Melkote, S., Liverani, E., Ascari, A., & Umbrello, D. (2018). Milling of maraging steel components produced by selective laser melting. *International Journal of Advanced Manufacturing Technology*, *94*(5–8), 1895–1902. https://doi.org/10.1007/s00170-017-0922-9
75. Kaynak, Y., & Tascioglu, E. (2018). Finish machining-induced surface roughness, microhardness and XRD analysis of selective laser melted Inconel 718 alloy. *Procedia CIRP*, *71*, 500–504. https://doi.org/10.1016/j.procir.2018.05.013
76. Kaynak, Y., & Kitay, O. (2019). The effect of post-processing operations on surface characteristics of 316L stainless steel produced by selective laser melting. *Additive Manufacturing*, *26*(December 2018), 84–93. https://doi.org/10.1016/j.addma.2018.12.021
77. Damon, J., Dietrich, S., Vollert, F., Gibmeier, J., & Schulze, V. (2018). Process dependent porosity and the influence of shot peening on porosity morphology regarding selective laser melted AlSi10Mg parts. *Additive Manufacturing*, *20*(December 2017), 77–89. https://doi.org/10.1016/j.addma.2018.01.001
78. Uzan, N. E., Ramati, S., Shneck, R., Frage, N., & Yeheskel, O. (2018). On the effect of shot-peening on fatigue resistance of AlSi10Mg specimens fabricated by additive manufacturing using selective laser melting (AM-SLM). *Additive Manufacturing*, *21*(September 2017), 458–464. https://doi.org/10.1016/J.ADDMA.2018.03.030
79. Baicheng, Z., Xiaohua, L., Jiaming, B., Junfeng, G., Pan, W., Chen-nan, S., … Jun, W. (2017). Study of selective laser melting (SLM) Inconel 718 part surface improvement by electrochemical polishing. *Materials and Design*, *116*, 531–537. https://doi.org/10.1016/j.matdes.2016.11.103
80. Zhang, B., Xiu, M., Tan, Y. T., Wei, J., & Wang, P. (2019). Pitting corrosion of SLM Inconel 718 sample under surface and heat treatments. *Applied Surface Science*, *490*, 556–567. https://doi.org/https://doi.org/10.1016/j.apsusc.2019.06.043

81. Brika, S. E., Letenneur, M., Dion, C. A., & Brailovski, V. (2020). Influence of particle morphology and size distribution on the powder flowability and laser powder bed fusion manufacturability of Ti-6Al-4V alloy. *Additive Manufacturing, 31*, 100929. https://doi.org/https://doi.org/10.1016/j.addma.2019.100929
82. Bai, Y., Chaudhari, A., & Wang, H. (2020). Investigation on the microstructure and machinability of ASTM A131 steel manufactured by directed energy deposition. *Journal of Materials Processing Technology, 276*, 116410.
83. Zhang, J., Chaudhari, A., & Wang, H. (2019). Surface quality and material removal in magnetic abrasive finishing of selective laser melted 316L stainless steel. *Journal of Manufacturing Processes, 45*, 710–719.
84. Wang, J., Zhu, J., & Liew, P. J. (2019). Material removal in ultrasonic abrasive polishing of additive manufactured components. *Applied Sciences, 9*(24), 5359.
85. Teng, X., Zhang, G., Zhao, Y., Cui, Y., Li, L., & Jiang, L. (2019). Study on magnetic abrasive finishing of AlSi10Mg alloy prepared by selective laser melting. *The International Journal of Advanced Manufacturing Technology, 105*(5), 2513–2521.
86. Han, S., Salvatore, F., Rech, J., Bajolet, J., & Courbon, J. (2020). Effect of abrasive flow machining (AFM) finish of selective laser melting (SLM) internal channels on fatigue performance. *Journal of Manufacturing Processes, 59*, 248–257.
87. Zhang, J., & Wang, H. (2019). Micro-blasting of 316L tubular lattice manufactured by laser powder bed fusion. In *Proceedings of the 19th International Conference of the European Society for Precision Engineering and Nanotechnology EUSPEN.*
88. Peng, X., Kong, L., Fuh, J. Y., & Wang, H. (2021). A Review of Post-Processing Technologies in additive manufacturing. *Journal of Manufacturing and Materials Processing*. https://doi.org/10.3390/jmmp5020038
89. Petters, R., Kühn, U., Löber, L., Flache, C., & Eckert, J. (2013). Comparison of different post processing technologies for SLM generated 316l steel parts. *Rapid Prototyping Journal, 19*(3), 173–179. https://doi.org/10.1108/13552541311312166
90. AlMangour, B., & Yang, J.-M. (2016). Improving the surface quality and mechanical properties by shot-peening of 17-4 stainless steel fabricated by additive manufacturing. *Materials & Design, 110*, 914–924.
91. Maamoun, A. H., Elbestawi, M., & Veldhuis, S. (2018). Influence of shot peening on AlSi10Mg parts fabricated by additive manufacturing. *Journal of Manufacturing and Materials Processing, 2*(3), 40. https://doi.org/10.3390/jmmp2030040
92. Bagherifard, S., Beretta, N., Monti, S., Riccio, M., Bandini, M., & Guagliano, M. (2018). On the fatigue strength enhancement of additive manufactured AlSi10Mg parts by mechanical and thermal post-processing. *Materials & Design, 145*, 28–41. https://doi.org/10.1016/j.matdes.2018.02.055
93. Kaynak, Y., & Tascioglu, E. (2020). Post-processing effects on the surface characteristics of Inconel 718 alloy fabricated by selective laser melting additive manufacturing. *Progress in Additive Manufacturing, 5*(2), 221–234. https://doi.org/10.1007/s40964-019-00099-1
94. Kaynak, Y., & Kitay, O. (2018). Porosity, surface quality, microhardness and microstructure of selective laser melted 316L stainless steel resulting from finish machining. *Journal of Manufacturing and Materials Processing, 2*(2), 36. https://doi.org/10.3390/jmmp2020036
95. Khan, H. M., Sirin, T. B., Tarakci, G., Bulduk, M. E., Coskun, M., Koc, E., & Kaynak, Y. (2021). Improving the surface quality and mechanical properties of selective laser sintered PA2200 components by the vibratory surface finishing process. *SN Applied Sciences, 3*(3), 364. https://doi.org/10.1007/s42452-021-04371-4

96. Sangid, M. D., Stori, J. A., & Ferriera, P. M. (2011). Process characterization of vibrostrengthening and application to fatigue enhancement of aluminum aerospace components—part I. Experimental study of process parameters. *The International Journal of Advanced Manufacturing Technology*, *53*(5), 545–560. https://doi.org/10.1007/s00170-010-2857-2
97. Canals, L., Badreddine, J., McGillivray, B., Miao, H. Y., & Levesque, M. (2019). Effect of vibratory peening on the sub-surface layer of aerospace materials Ti-6Al-4V and E-16NiCrMo13. *Journal of Materials Processing Technology*, *264*, 91–106. https://doi.org/10.1016/j.jmatprotec.2018.08.023
98. Kolganova, E. N., Goncharov, V. M., & Fedorov, A. V. (2019). Investigation of deburring process at vibro-abrasive treatment of parts having small grooves and holes. *Materials Today: Proceedings*, *19*, 2368–2373. https://doi.org/10.1016/j.matpr.2019.07.726
99. Kang, Y.-J., Yang, S., Kim, Y.-K., AlMangour, B., & Lee, K.-A. (2019). Effect of post-treatment on the microstructure and high-temperature oxidation behaviour of additively manufactured inconel 718 alloy. *Corrosion Science*, *158*, 108082. https://doi.org/10.1016/j.corsci.2019.06.030
100. Ma, C. P., Guan, Y. C., & Zhou, W. (2017). Laser polishing of additive manufactured Ti alloys. *Optics and Lasers in Engineering*, *93*, 171–177. https://doi.org/10.1016/j.optlaseng.2017.02.005
101. Kruth, J.-P., Deckers, J., & Yasa, E. (2008). Experimental investigation of laser surface remelting for the improvement of selective laser melting process. In *2008 International Solid Freeform Fabrication Symposium*.
102. Lamikiz, A., Sánchez, J. A., López de Lacalle, L. N., & Arana, J. L. (2007). Laser polishing of parts built up by selective laser sintering. *International Journal of Machine Tools and Manufacture*, *47*(12–13), 2040–2050. https://doi.org/10.1016/j.ijmachtools.2007.01.013
103. Mai, T. A., & Lim, G. C. (2004). Micromelting and its effects on surface topography and properties in laser polishing of stainless steel. *Journal of Laser Applications*, *16*(4), 221–228.
104. Pfefferkorn, F. E., Duffie, N. A., Morrow, J. D., & Wang, Q. (2014). Effect of beam diameter on pulsed laser polishing of S7 tool steel. *CIRP Annals*, *63*(1), 237–240. https://doi.org/10.1016/j.cirp.2014.03.055
105. Chow, M. T. C., Bordatchev, E. V, & Knopf, G. K. (2013). Experimental study on the effect of varying focal offset distance on laser micropolished surfaces. *The International Journal of Advanced Manufacturing Technology*, *67*(9), 2607–2617. https://doi.org/10.1007/s00170-012-4677-z
106. Bagherifard, S., Slawik, S., Fernández-Pariente, I., Pauly, C., Mücklich, F., & Guagliano, M. (2016). Nanoscale surface modification of AISI 316L stainless steel by severe shot peening. *Materials & Design*, *102*, 68–77. https://doi.org/10.1016/j.matdes.2016.03.162
107. Kempen, K., Thijs, L., Van Humbeeck, J., Kruth, J.-P. P., & K. Kempen, L. T. J. V. H. J.-P. K. (2012). Mechanical properties of AlSi10Mg produced by selective laser melting. *Physics Procedia*, *39*(March 2017), 439–446. https://doi.org/10.1016/j.phpro.2012.10.059
108. Read, N., Wang, W., Essa, K., & Attallah, M. M. (2015). Selective laser melting of AlSi10Mg alloy: Process optimisation and mechanical properties development. *Materials & Design (1980-2015)*, *65*, 417–424. http://dx.doi.org/10.1016/j.matdes.2014.09.044

109. Maskery, I., Aboulkhair, N. T., Aremu, A. O., Tuck, C. J., & Ashcroft, I. A. (2017). Compressive failure modes and energy absorption in additively manufactured double gyroid lattices. *Additive Manufacturing*, *16*, 24–29. https://doi.org/10.1016/j.addma.2017.04.003
110. Hitzler, L., Charles, A., & Öchsner, A. (2016). The influence of post-heat-treatments on the tensile strength and surface hardness of selective laser melted AlSi10Mg. In *Defect and Diffusion Forum* (Vol. 370, pp. 171–176). Trans Tech Publ.
111. Solberg, K., Hovig, E. W., Sørby, K., & Berto, F. (2021). Directional fatigue behaviour of maraging steel grade 300 produced by laser powder bed fusion. *International Journal of Fatigue*, *149*, 106229.
112. Wei, S., Kumar, P., Lau, K. B., Wuu, D., Liew, L.-L., Wei, F., … Ramamurty, U. (2022). Effect of heat treatment on the microstructure and mechanical properties of 2.4 GPa grade maraging steel fabricated by laser powder bed fusion. *Additive Manufacturing*, *59*, 103190. https://doi.org/https://doi.org/10.1016/j.addma.2022.103190
113. Jägle, E. A., Choi, P.-P., Van Humbeeck, J., & Raabe, D. (2014). Precipitation and austenite reversion behavior of a maraging steel produced by selective laser melting. *Journal of Materials Research*, *29*(17), 2072–2079. https://doi.org/10.1557/jmr.2014.204
114. Cyr, E., Asgari, H., Shamsdini, S., Purdy, M., Hosseinkhani, K., & Mohammadi, M. (2018). Fracture behaviour of additively manufactured MS1-H13 hybrid hard steels. *Materials Letters*, *212*, 174–177. https://doi.org/10.1016/j.matlet.2017.10.097
115. Wu, P. Y., & Yamaguchi, H. (2018). Material removal mechanism of additively manufactured components finished using magnetic abrasive finishing. *Procedia Manufacturing*, *26*, 394–402. https://doi.org/10.1016/j.promfg.2018.07.047
116. Mercelis, P., & Kruth, J.-P. (2006). Residual stresses in selective laser sintering and selective laser melting. *Rapid Prototyping Journal*, *12*(5), 254–265.
117. AlMangour, B., Grzesiak, D., & Yang, J.-M. (2017). Selective laser melting of TiB2/316L stainless steel composites: The roles of powder preparation and hot isostatic pressing post-treatment. *Powder technology*, *309*, 37–48.
118. Dumas, M., Cabanettes, F., Kaminski, R., Valiorgue, F., Picot, E., Lefebvre, F., … Rech, J. (2018). Influence of the finish cutting operations on the fatigue performance of Ti-6Al-4V parts produced by selective laser melting. In *Procedia CIRP* (Vol. 71, pp. 429–434). Elsevier. https://doi.org/10.1016/j.procir.2018.05.054
119. Vandenbroucke, B., & Kruth, J.-P. J. (2007). Selective laser melting of biocompatible metals for rapid manufacturing of medical parts. *Rapid Prototyping Journal*, *13*(4), 196–203.
120. Facchini, L., Magalini, E., Robotti, P., Molinari, A., Höges, S., & Wissenbach, K. (2010). Ductility of a Ti-6Al-4V alloy produced by selective laser melting of prealloyed powders. *Rapid Prototyping Journal*, *16*(6), 450–459. https://doi.org/10.1108/13552541011083371
121. Craeghs, T., Thijs, L., Verhaeghe, F., Kruth, J.-P., & Humbeeck, J. Van. (2010). A study of the microstructural evolution during selective laser melting of Ti–6Al–4V. *Acta Materialia*, *58*(9), 3303–3312. https://doi.org/10.1016/j.actamat.2010.02.004
122. Demir, A. G., & Previtali, B. (2017). Additive manufacturing of cardiovascular CoCr stents by selective laser melting. *Materials & Design*, *119*, 338–350. https://doi.org/10.1016/j.matdes.2017.01.091
123. Yung, K. C., Wang, W. J., Xiao, T. Y., Choy, H. S., Mo, X. Y., Zhang, S. S., & Cai, Z. X. (2018). Laser polishing of additive manufactured CoCr components for controlling their wettability characteristics. *Surface and Coatings Technology*, *351*, 89–98. https://doi.org/10.1016/j.surfcoat.2018.07.030

124. Sing, S. L., Huang, S., & Yeong, W. Y. (2020). Effect of solution heat treatment on microstructure and mechanical properties of laser powder bed fusion produced cobalt-28chromium-6molybdenum. *Materials Science and Engineering: A, 769*, 138511. https://doi.org/10.1016/j.msea.2019.138511
125. Ayyıldız, S., Soylu, E. H., İde, S., Kılıç, S., Sipahi, C., Pişkin, B., & Gökçe, H. S. (2013). Annealing of Co-Cr dental alloy: effects on nanostructure and Rockwell hardness. *The Journal of Advanced Prosthodontics*, *5*(4), 471–478.
126. Careri, F., Imbrogno, S., Umbrello, D., Attallah, M. M., Outeiro, J., & Batista, A. C. (2021). Machining and heat treatment as post-processing strategies for Ni-superalloys structures fabricated using direct energy deposition. *Journal of Manufacturing Processes*, *61*, 236–244.
127. Özer, G., Tarakçi, G., Yilmaz, M. S., Öter, Z., Sürmen, Ö., Akça, Y., … Koç, E. (2020). Investigation of the effects of different heat treatment parameters on the corrosion and mechanical properties of the AlSi10Mg alloy produced with direct metal laser sintering. *Materials and Corrosion*, *71*(3), 365–373. https://doi.org/10.1002/maco.201911171
128. Özer, G., & Karaaslan, A. (2020). A study on the effects of different heat-treatment parameters on microstructure–mechanical properties and corrosion behavior of maraging steel produced by direct metal laser sintering. *Steel Research International*, *91*(10), 1–8. https://doi.org/10.1002/srin.202000195
129. Özer, G. (2022). Heat treatment, microstructure and corrosion relationship on 316L SS produced by AM. *Materials Science and Technology*, 1–12. https://doi.org/10.1080/02670836.2022.2131242
130. Kong, D., Dong, C., Ni, X., Zhang, L., Yao, J., Man, C., … Li, X. (2019). Mechanical properties and corrosion behavior of selective laser melted 316L stainless steel after different heat treatment processes. *Journal of Materials Science & Technology*, *35*(7), 1499–1507. https://doi.org/10.1016/j.jmst.2019.03.003
131. Guzanová, A., Draganovská, D., Ižaríková, G., Živčák, J., Hudák, R., Brezinová, J., & Moro, R. (2019). The effect of position of materials on a build platform on the hardness, roughness, and corrosion resistance of Ti6Al4V produced by DMLS technology. *Metals*. https://doi.org/10.3390/met9101055

5 Thermal Post-Processing Techniques for Additive Manufacturing

Muhammad Arif Mahmood[1], *Abid Ullah*[2], *Mussadiq Shah*[3], *Asif Ur Rehman*[4], *Metin Uymaz Salamci*[5], *and Marwan Khraisheh*[6]

[1]Intelligent Systems Center, Missouri University of Science and Technology, Rolla, MO, USA

[2]Additive Manufacturing Technologies Application and Research Center-EKTAM, Gazi University, Ankara, Turkey

[3]Mechanical Engineering Department, Gazi University, Ankara, Turkey; Additive Manufacturing Technologies Research and Application Center-EKTAM, Gazi University, Ankara, Turkey

[4]ERMAKSAN, Bursa, Turkey; Mechanical Engineering Department, Gazi University, Ankara, Turkey; Additive Manufacturing Technologies Research and Application Center-EKTAM, Gazi University, Ankara, Turkey

[5]Mechanical Engineering Department, Gazi University, Ankara, Turkey; Additive Manufacturing Technologies Research and Application Center-EKTAM, Gazi University, Ankara, Turkey

[6]Mechanical Engineering Program, Texas A&M University at Qatar, Qatar

CONTENTS

DOI: 10.1201/9781003276111-5

5.1 INTRODUCTION

Additive manufacturing (AM) can produce three-dimensional (3D) objects layer-wise based on the CAD model (Chen et al. 2021). AM has seen substantial modifications in its production concept, raw materials, and part performance (Pandey, Reddy, and Dhande 2003). Due to layer-after-layer printing, AM can quickly build intricate 3D structural parts (Dai et al. 2019). Modifying the non-equilibrium solidification procedure allows the components' development with tailor-made qualities (Pyka et al. 2013). The advantages of AM include a shorter manufacturing cycle and lower per-unit costs when produced in small batches (Charalampous, Kostavelis, and Tzovaras 2020). In contrast to traditional manufacturing methods, this method only requires raw materials and equipment to manufacture parts without complex tooling. Other benefits of AM methods include near-net-part development, minimal machining allowance, and high material usage (Ko, Moon, and Hwang 2015). As a result of AM's elevated energy delivered to a given area, different materials can be processed (Thompson, Maskery, and Leach 2016). The energy provided during printing efficiently raises the temperature in a local area to a very high value, sufficient to liquefy metallic materials. When opposed to casting, the AM metallic pieces have significant residual stresses (Taheri et al. 2017). The forming stress, also known as residual stress, is released when a layer is deposited by 3D printing and allowed to solidify. Local heating and cooling of top surfaces and re-melting of the preceding layers are essential steps in laser-based-AM processes. Residual stresses are produced due to inhomogeneous thermal loads. The temperature gradient mechanism (TGM) and cool-down mechanism (CDM), describing the melting, solidification, and re-melting processes, is the primary source of residual stresses (Mahmood et al. 2020). In TGM, when a heating source with a high intensity interacts with the material, the material's temperature rises quickly in the vicinity of the heat source compared to the rest of the material. However, the irradiated regime is constrained by the cooler (room-temperature) material, resulting in "compressive stress." When the material cools down to room temperature as the laser beam moves away, the material's contraction occurs that is constrained by the surrounding material, resulting in "tensile residual stresses," known as CDM (Carpenter and Tabei 2020). The re-melting of the previously deposited layer also contributes to residual stress formation. When a new layer is deposited by melting the powder particles, the previously deposited layer experiences the re-melt or near re-melt due to localized high temperature, which not only affects the residual stresses magnitude in the previous layer cools but also in the new layer. "Lattice spacing" inhomogeneity caused by the inhomogeneous microstructure formation is another source of location-dependent residual stresses, which can be attributed to the non-equilibrium thermal loads in the AM (Carpenter and Tabei 2020).

AM methods have several benefits, such as producing complex structures with various materials, including composites with high processing efficiency (Kong et al. 2020). However, a major concern with AM methods is that the surface quality of AM products is often lower than that of conventional manufacturing procedures due to the layer-by-layer deposition process (Zhang et al. 2018). The roughness of surfaces can vary considerably between various AM techniques. Consequently, AM cannot produce products that simultaneously meet mechanical and surface roughness specifications (Tan et al. 2020). The influential factor is the absence of physics understanding involved in the printing process, which provides difficulty in explaining the melting process for laser additive manufacturing (LAM). For instance, when assessing selective laser melting (SLM), one must consider the rapid solidification phenomena that occurs in a very high thermal gradient and the extremely strong bonding force in processing zones. A deeper understanding of how the internal structure and thermal stress of these components change with repeated heating and cooling is required. Balling causes, pores' development, cracks' initiation, powder particles' accumulation, and thermally induced stresses that commonly occur during AM. A manufactured part's structural mechanics can be severely compromised by these defects (Cerniglia and Montinaro 2018). As a result, printed objects need post-processes to enhance the thermophysical characteristics before operational utilization (Chavez et al. 2020).

This chapter discusses the various defects in the AM parts, including porosity, cracking, anisotropy, and surface roughness, and their elimination using thermal post-process methods, such as laser shock peening and polishing, hot-isostatic pressing, and annealing.

5.2 DEFECTS IN ADDITIVELY MANUFACTURED PARTS

In metallic, mostly traditional metallic materials are used as feedstocks for AM. Defects involved in the AM methods are usually deprived surface roughness, cracks' evolution, inhomogeneity of the mechanical characteristics, and porosity (Aboulkhair et al. 2019). There are numerous underlying causes of defect creation in the AM process. Here, we have listed a number of flaws and their reasons.

5.2.1 Surface Roughness

Most components manufactured by the AM technique have poor surface roughness (SR). In contrast, the ideal approach is to manufacture parts using AM without needing post-processing. This objective has not been achieved yet. One of the ways to resolve this issue is by developing a one-to-one correlation between inputs and outputs. However, increasing the SR properties may introduce new faults into the manufactured components, limiting the number of potential options. The laser parameters employed directly impact the melt pool's stability and, consequently, the bead's uniformity (Kempen et al. 2011). Additionally, AM exhibits the balling phenomenon, resulting in a coarse melt pool that affects the resultant SR. The operating conditions control the printed surface morphology and SR (Olakanmi 2013). In utilized conditions, laser scanning speed significantly impacts SR, such that the low laser scanning speed causes SR elevation (Louvis, Fox, and Sutcliffe 2011).

5.2.2 Crack Formation

Crack formation is another flaw observed during the AM (Rappaz, Drezet, and Gremaud 1999b). Classical casting or welding processes can better understand the failure mechanisms of metals with poor processability during AM. Certain metals are prone to cracking during solidification, including stainless steels and Al- and Ni-based alloys (Kou 2015). The solidification shrinkage and thermal contraction during cooling are the primary causes of these cracks. In addition, crack formation in the semi-solid area is aided by thermal stresses caused by the fabrication process's nature. This fracture type is characterized as hot-tearing in casting (Campbell 2004) and solidification cracking in welding ("Library" 2003). When the flow of liquid material is insufficient, the tensile stresses in the mushy zone generate the hot-tears, as proposed by the hot-tearing model for metallic alloys (Rappaz, Drezet, and Gremaud 1999a). Intergranular solidification cracks were discovered in Co-Cr-Fe-Ni alloy (Sun et al. 2019). Since no elemental segregations could be found at grain boundaries, the substantial residual stresses induced by the coarse grains cause the formation of cracks. Cracks typically initiate during solidification and grow due to heat cycling in the solid state (Tomus et al. 2017). Hot-tearing and solidification cracking are both problems for AM parts. The degree to which this vulnerability manifests depends on several elements, including grain size, residual stresses, grain boundary segregations, and liquid films on grain borders during solidification.

5.2.3 Inhomogeneity of Mechanical Properties

The inhomogeneity of mechanical properties is another form of defect in AM process. The degree of anisotropy is subjected to the "orientation" used in part printing and the complex cyclic-heat generation throughout the printing process (Dai et al. 2017). The inhomogeneity of the mechanical properties can be controlled by optimizing the laser scanning strategy (Thijs et al. 2013). The building orientation and utilized supports are critical in controlling the inhomogeneity concerning the thermophysical characteristics of the printed components (Kimura and Nakamoto 2016).

5.2.4 Porosity

Porosity is a commonly known defect in printed components (Kruth et al. 2010). The performance of a part is directly related to the printed component density, thus defining its strength (Morgan 2014). Porosity may be developed by the entrapment of gases or the evaporation of an element during printing, as well as by lack of fusion that may result in catastrophes (Frazier 2014). The porosities can be divided into macro- and microporosities. Sinico et al. (2021) quantified macro- and microporosities based on the diameter in the case of Cu-Cr1 alloy. They concluded that macropores usually have a diameter between 26–145 μm, while the micropores' diameter range is 21–93 μm. Besides, the macroporosities result from a lack of fusion and are not a sphere in aspect ratio considering a circular profile. Mostly they contribute to fracture generation and propagation. Microporosities may become crucial after heat treatment (Li et al. 2016). Microporosities tend to consolidate, presenting macropores (Galy et al. 2018). Porosity can be accredited to numerous factors, including operating conditions, the existence of impurities in the material to be printed, the reduced laser absorptivity of the material, wettability issues, the printing chamber conditions, and evaporation of the alloying elements (Kimura and Nakamoto 2016).

Various disadvantages associated with AM techniques hinder their implementation in demanding industries. However, various methods enhance the structural and mechanical qualities of the AM components. In the following section, a few of the strategies are described below.

5.3 THERMAL POST-PROCESSING TECHNIQUES

This section compiles the various thermal techniques for AM parts.

5.3.1 Laser Shock Peening

Laser shock peening involves the compression of the material at the plastic level. When laser shock peening is performed, the given material's ability to bear the transverse strain causes the accumulation of local compressive stresses (Raja et al. 2018). Laser shock peening usually causes compressive stress-strain phenomena for a given material. However, the induced stress-strain is substantially greater in thin parts than in thick components (Velu et al. 2021). The schematic of the laser shock peening process can be identified in Figure 5.1, performed on a metallic part. Using a concentrated laser beam for about 30 ns, the heated zone urges to 10^4 °C, which leads to "plasma" generation. This plasma captivates the laser energy significantly. The pressure made by plasma is transmitted to the substantial by shock waves (Sundar et al. 2019). Laser shock peening involves a "confined-mode" that is often used to get a high shock pressure amplitude. The metallic material is usually layered with a non-transparent material and is protected against direct laser light. Recent research shows that when the confined mode is used, the pressure of the plasma on the metal surface can reach up to 10 GPa. When the amount of compressive residual stress-strain is high, a stronger pressure pulsation generates the pressure to a deeper depth (Ding and Ye 2006).

Laser shock peening is typically utilized to increase the durability of a part in case aircraft parts last longer (Hackel et al. 2018). Laser shock peening was applied to increase the fatigue strength of the steels (Lan et al. 2020). Short and intense pulses generate plasma for a given part, which in-turn increases the pressure, causing deformation at the local level. The enhanced, developed pressure results in an effective operation (Hackel et al. 2018). In laser shock peening stress waves are generated, which

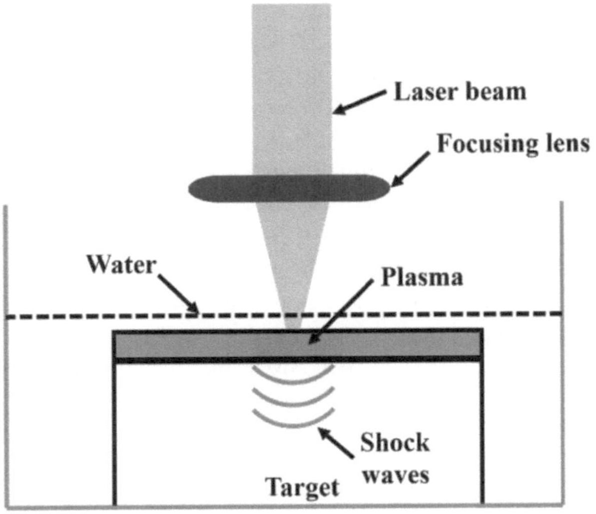

FIGURE 5.1 Schematic of laser shock peening; drawn based on the data provided in Ref. (Ding and Ye 2006).

change the microstructure along with corresponding mechanical properties. In the case of Al, the yield strength of 7075-Al was increased by 30% compared to the un-shocked samples (Fairand et al. 2003).

The effect of laser shock peening on Al-6061-T6 was investigated. The operating conditions, with a laser beam (Nd: YAG) having 1200 mJ energy and 8 ns pulse duration. (Salimianrizi et al. 2016). The results indicated that compressive residual stress-strain was generated on the processed surface. In addition, surface roughness studies revealed that, depending on the operating conditions, the LSP could decrease surface quality due to localized deformation. A parametric study for Inconel 718 by AM was conducted by varying the laser power and pulse number (Jinoop et al. 2019). The sample hardness and depth were measured with laser power = 170 mW, and the number of laser-material interaction shots = 7. It was discovered that LSP affected the surface quality and corresponding mechanical characteristics of the manufactured specimens. The analysis discovered that the Vickers microhardness equivalent to 360 HV at 10 μm depth was determined. After treatment, the compressive residual stress-strain in the manufactured parts increased from 214.9 to 307.9 MPa. The wear rate also increases up to 1.7 times more than untreated specimens. The wearing characteristics of both types of samples are compiled in Figure 5.2(a)–(e) (Jinoop et al. 2019). It was also revealed that laser shock peening lessens the debris particles present on a surface due to an enhanced residual stress-strain. It can be attributed to the decreased number of holes in laser shock peening-treated samples over untreated samples. As previously described, a material subjected to laser shock peening undergoes melting and solidification at the local level, which minimizes the pores. Figure 5.2(e) illustrates the variance in wearing rate due to various laser shock peening parameters. It was revealed that the wearing rate decreases by increasing the

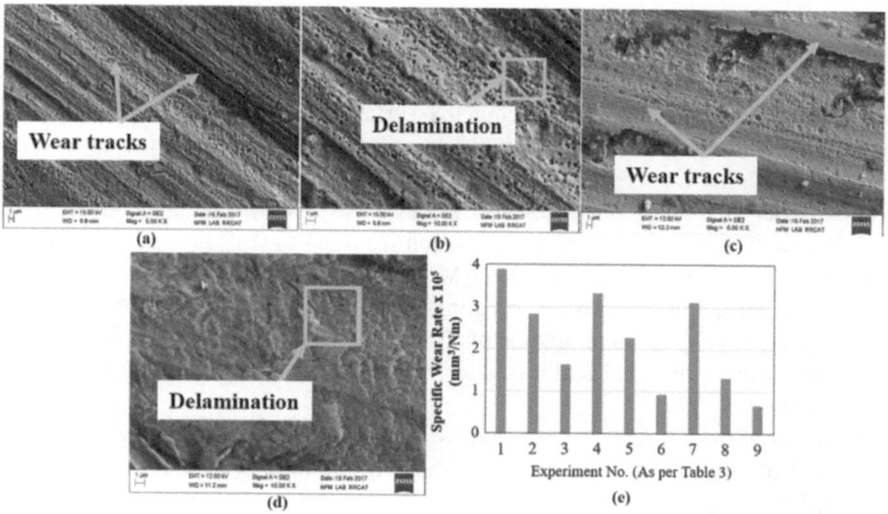

FIGURE 5.2 Wearing rate (a, b) for un-treated laser shock peening samples, (c, d) for treated LSP samples, and (e) its comparison with untreated samples (Jinoop et al. 2019); reproduced after permission from Springer.

power and the shot number. Variation in laser power appears to be less significant than the number of shots, indicating that the lowermost wearing rate occurred at laser power = 200 mW with seven shots.

The effect of laser shock peening on In-718 AM specimens was investigated (Sidhu et al. 2019). After laser shock peening treatment, it was discovered that the residual stresses and hardness improved with increasing laser energy density such that the treated specimen exhibited a residual stress =–875 MPa (compressive) and hardness = 858 HV.

5.3.2 Laser Polishing

Laser polishing is usually applied to enhance the surface quality of AM components (Wang et al. 2015). During laser polishing, the morphological peaks rapidly reach the melting point when a surface is irradiated via a laser beam. After the melt pool is produced, the liquefied material reorganizes to the same level due to "gravity" and "surface tension." The temperature of the heat-affected region decreases as the beam travels away, resulting in the solidification of the melt pool and smoothing of the surface. Laser polishing is a re-melting process that modifies the surface morphology without changing the bulk characteristics (Ma, Guan, and Zhou 2017). Laser polishing was utilized to polish the SS-304 (Mai and Lim 2004). The melting depth was sub-microscopic, and the polishing rate was between 5 and 15 cm^2/m. The surface quality in terms of roughness was reduced from 195 to 75 nm because of laser polishing, increasing the 14% specular surface reflectance and decreasing the 70% diffuse surface reflectance. The heterogeneous distribution of microhardness was transformed into a homogeneous distribution. Due to the microstructural alterations caused by melting and solidification, laser polishing could exhibit enhanced resistance to pitting corrosion. The melting process is advantageous for sealing pores and cracks at microlevel, hence improving the surface quality. An orthogonal design experiment was developed to monitor the laser beam working conditions to shorten the investigational time required to achieve a superior surface finish on tool steel using a pulsed Nd: YAG laser (Guo et al. 2011). Based on the optimal conditions, the surface quality in terms of roughness reduced from 0.4 to 0.12 μm. Laser polishing was applied to the SS-420, and bronze AM parts (Lamikiz et al. 2007). It was found that the surface quality in terms of roughness reduced from 7.5 to 1.2 μm after laser polishing. According to metallurgical analyses, the heat-affected region included no pores. In contrast, the final surfaces were harder and more uniform than the basic components.

Titanium alloys by AM were treated by fiber laser (Ma, Guan, and Zhou 2017). Figure 5.3(a) and (e) depict macrophotographs of titanium alloy surfaces following laser polishing. As shown in Figure 5.3(b) and (f), the samples were polished. After laser polishing, the surface roughness of titanium alloy was reduced significantly, as depicted in Figure 5.3(c)–(d) and 5.3(g)–(h).

Laser polishing was utilized on titanium alloys to determine the optimal laser energy density for minimizing surface roughness (Zhou et al. 2019). The operating conditions were applied, including power = 150 W, scanning speed = 20 mm/s, and overlapping = 95%. The surface quality was improved by reducing the roughness from 3.09 to 0.56 μm. As depicted in Figure 5.4, the appearance of martensitic structures in the material resulted from cyclic heating and cooling, leading to a 25% hardness improvement.

Before and after the laser polishing process, the morphology and microstructure of SS-316L components were studied (Chen et al. 2021). After laser polishing, the surface roughness diminished from 4.84 to 0.65 μm, and hardness equivalent to 262 HV was attained. The effect of LP on the surface roughness of SS-316L was studied (Rosa, Mognol, and Hascoët 2015). After five laser passes, the surface roughness decreased by 96%. Utilizing a CW fiber laser, the surface of $\alpha+\beta$ titanium alloy was treated (Lee et al. 2021). A unique surface was obtained by re-melting the powder particles with laser polishing. The outcomes are depicted in Figure 5.5(a) and (b). It can be observed that the surface of the as-built samples is incredibly random. The surfaces following laser polishing are shown in Figure 5.5(c) and (d). One may note that laser polishing yielded surfaces with small peak-to-valley aspect ratios and smooth topographies.

Laser polishing also affects the durability of the treated specimens (Avilés et al. 2011) as it was applied to AISI-1045 steel. It was identified that laser polishing increased the durability up to 10^6 more compared to the un-treated samples.

5.3.3 Hot Isostatic Pressing

Hot isostatic pressing (HIP) is the application of temperature and pressure to as-built components for a specified time. Temperature and pressure are the two critical parameters in HIP pressing, with samples kept in an argon environment to minimize

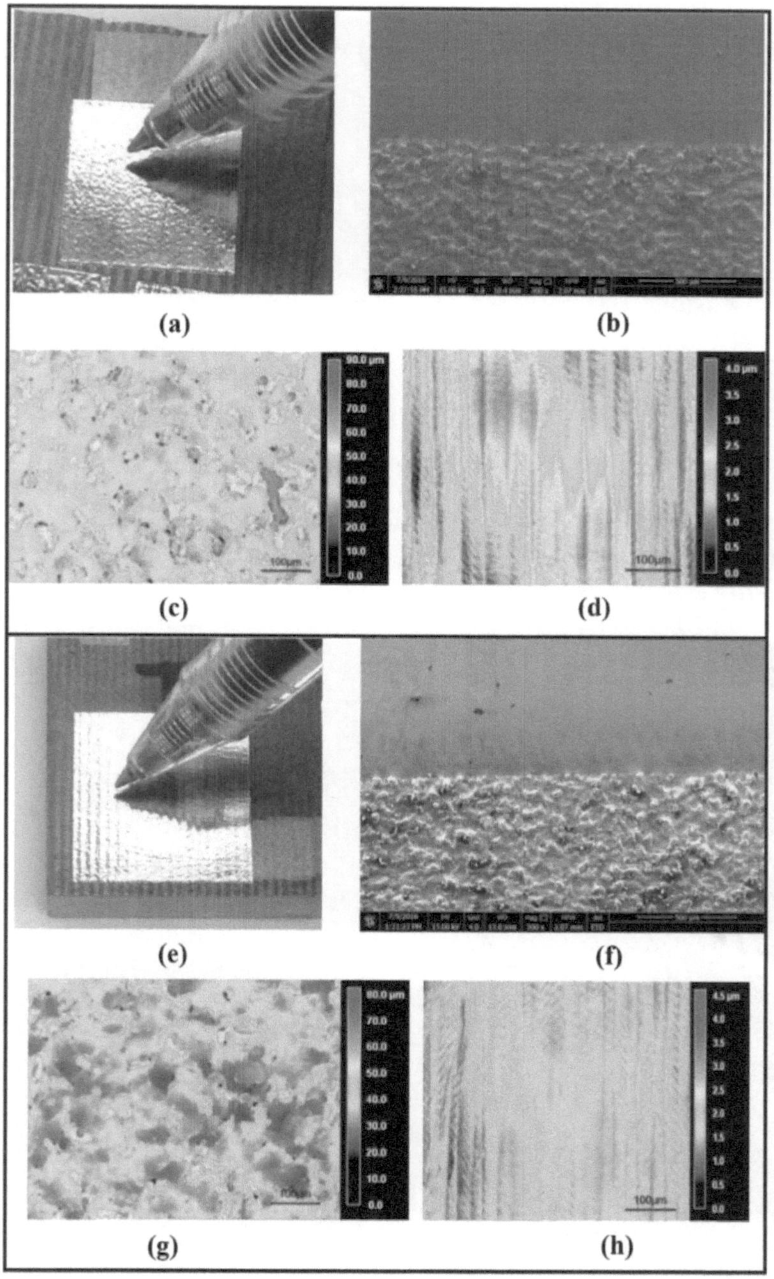

FIGURE 5.3 Ti6Al4V (a) laser polished (LP) specimen, (b) scanning electron microscopy after LP, (c) as-fabricated specimen topography, (d) after LP topography, and $\alpha + \beta$ Ti-alloy (e) LP specimen, (f) scanning electron microscopy after LP, (g) as-fabricated specimen topography, (h) after LP topography (C. P. Ma, Guan, and Zhou 2017); reproduced after permission from Elsevier.

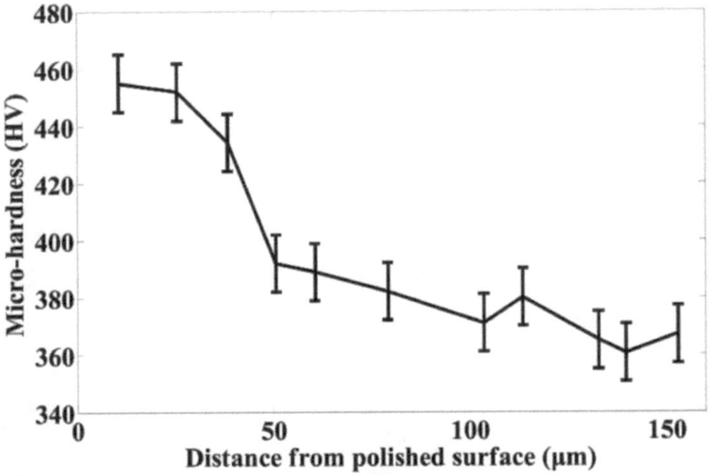

FIGURE 5.4 Correlation between polished surface distance and hardness (Zhou et al. 2019); reproduced after permission from Elsevier.

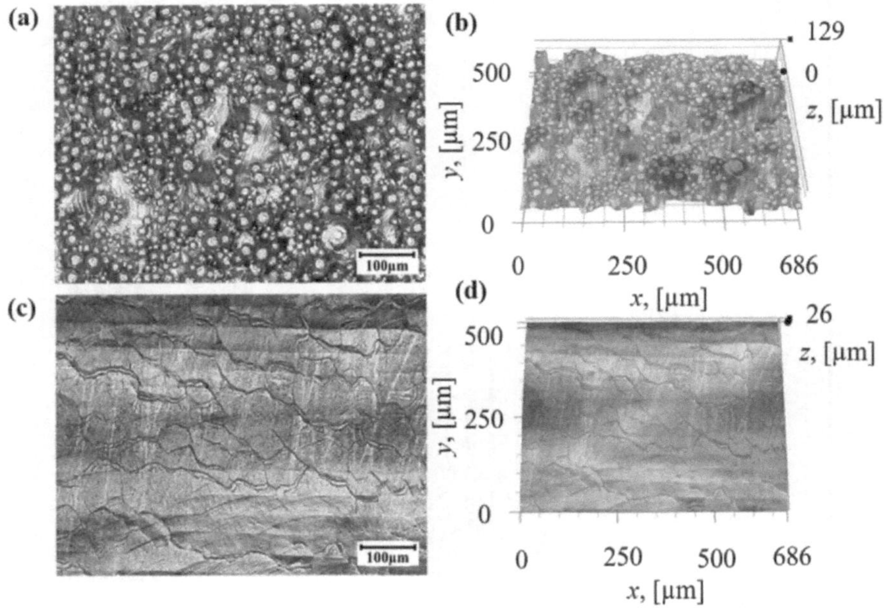

FIGURE 5.5 (a, b) Morphology and topography of α + β titanium alloy prior to laser polish, and (c, d) morphology and topography of α + β titanium alloy after laser polish (Lee et al. 2021); reproduced after permission from Elsevier.

oxidation. HIP is one of the major post-processing approaches used for eliminating part defects, such as lack-of-fusion, keyhole, surface roughness, residual stress, and gas pores. Metal assemblies that previously needed multiple pieces may now be produced as one part by combining HIP with AM, which results in considerable production cost savings. Previous studies have primarily employed the HIP approach to reduce porosity, alleviate anisotropy, relieve stresses, and obtain the best mechanical properties in AM parts, which has shown significant results (Herzog et al. 2016; Shiyas and Ramanujam 2021; Chen et al. 2019; Lavery et al. 2017). However, if the temperatures are too high, mechanical strength and corrosion resistance can be reduced (Grech et al. 2022). HIP processes often use an inert gas at high pressures to produce an isostatic force on a material at high temperatures. This approach may minimize porosity and heat treat the product in a single step, resulting in enhanced mechanical properties. The intense pressure associated with HIP treatments improves the solubility of gases trapped inside pores; these gases then move to the part's surface, where the internal pore shrinks and fills (Atkinson and Davies 2000; Grech et al. 2022). However, due to the invasion of the HIP treatment gas into the porosity during the process, HIP treatments may not be able to remove porosity that is coupled to a part's exterior surface (Du Plessis and Macdonald 2020; Cegan et al. 2020). This signifies that additional machining may also be required to eliminate any remaining surface-breaking porosity.

Defects such as gas pores, lack of fusion, morphology-related large pores, keyhole, and partly melted powders are common in laser powder bed fusion (LPBF) fabricated metals and alloys. These defects degrade fatigue qualities and may interfere with LPBF process quality control. Numerous studies have been conducted to investigate the impact of HIP treatments on eliminating porosity and how it may regulate improvements to the microstructure and subsequent mechanical characteristics of metal AM (Cao et al. 2021; Cegan et al. 2020). HIP has been proven to be an effective method for enhancing the fatigue performance of additively manufactured alloy samples by closing the internal pores (Puichaud et al. 2019). Plessis et al. (Du Plessis and Macdonald 2020) investigated the impact of HIP on the keyholes and porosity reduction in LPBF-manufactured Ti6Al4V. They reported that the HIP process could assist in consolidating un-melted or partly melted powder particles and minimize porosity and keyholes. Figure 5.6(a) and (b) show that the sample obtained at 120 W laser power contains 0.6% porosity randomly dispersed before HIP, and substantially all of these pores are closed during HIP. Similarly, the sample has a significant number of keyhole pores, which were significantly reduced by the HIP process, and nearly no keyholes remain in the middle of the manufactured part, as shown in Figure 5.6(c) and (d). A similar effect was observed in the electron beam melting Ti-6Al-4V, LPBF 17–4PH steel, LPBF Inconel 718, and LPBF 316L, where the HIP process essentially reduced the internal porosity and significantly enhanced the fatigue properties of the parts compared to the as-printed (Liu et al. 2022; Grech et al. 2022).

Controlled porosity following the HIP process improves the fabricated part's surface quality and mechanical performance (Cegan et al. 2020). For example, the ductility of the Ti6Al4V component produced by LPBF improved to 15–20% following

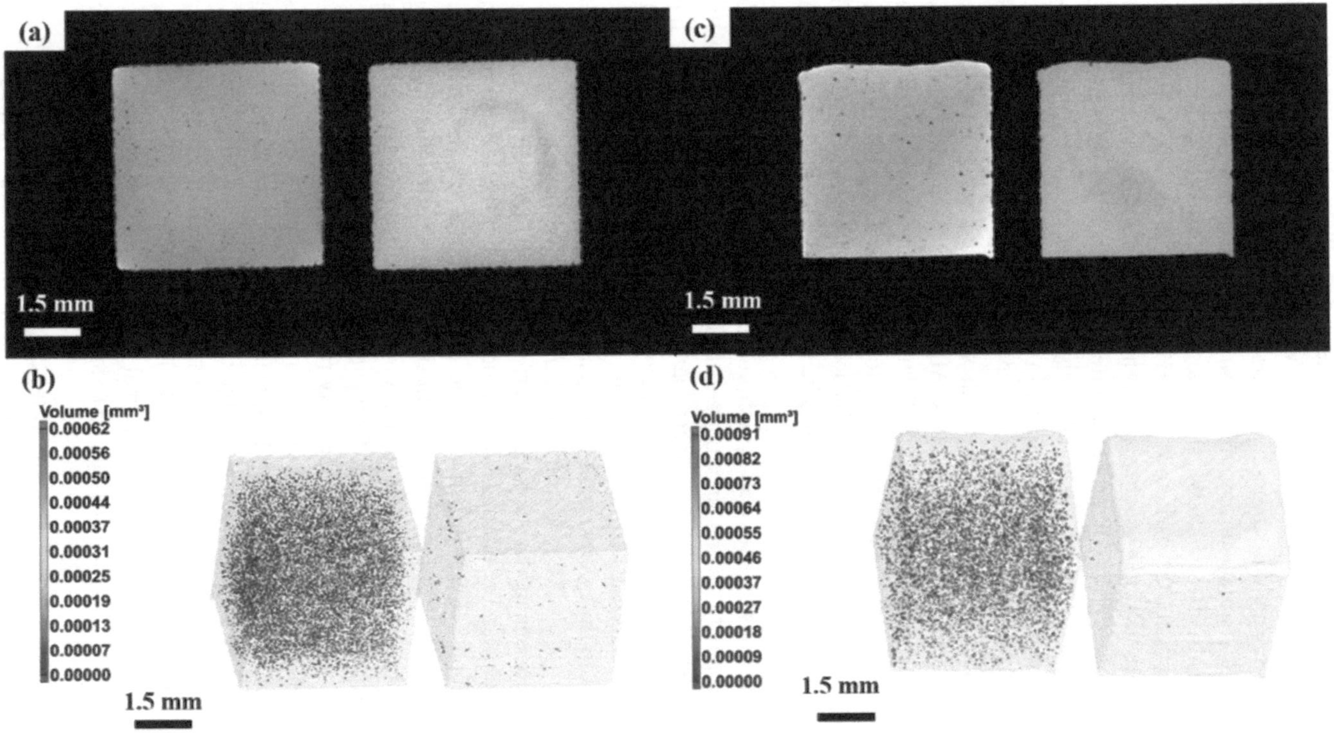

FIGURE 5.6 The influence of HIP treatment on the closure of defects in laser powder bed fusion-fabricated Ti-6Al-4V (a, b), and porosity before and after HIP process (left to right). (c, d) Keyholes before and after HIP process (left to right) (Du Plessis and Macdonald 2020); reproduced after permission from Elsevier.

HIT compared to 5–10% as-printed, with just a slight reduction in the yield strength (Zerbst and Madia 2021). Previous research indicates that temperature changes significantly impact the component's microstructure more during the HIP process than pressure variations (Liverani et al. 2020). Grech et al. (2022) looked at the best cycles for laser powder bed fusion-produced stainless steel 316L with the best mechanical characteristics (tensile, density, hardness, and low cycle fatigue), as well as pitting corrosion resistance. Melt pools, melt pool borders, and sub-grains cellular characteristics present in the microstructure and influence the mechanical performance of the finished part were removed during post-processing HIP cycles at 1125–1200 °C. However, they explored that the HIP process carried out at the lowest temperature (700 °C) led to an increase in yield strength and a decrease in elongation to fracture. Sripada et al. (2022) reported promising improvements in the microstructure with higher mechanical characteristics of the additively manufactured 17-4PH steel. The decreased defect density and densification caused by HIP enhanced the mechanical properties. Figure 5.7 depicts the compressive true stress-strain curves of LPBF-fabricated as-printed and HIP specimens. Zhai et al. (2019) studied the effect of the HIP on the microstructure and mechanical characteristics of CLAM steel produced by the LPBF process. They identified considerable microstructural improvement, including reduced microdefects in the fabricated parts by subjecting them to the HIP process. They revealed that dislocation migration and recrystallization during the HIP process enhanced the microstructure and mechanical characteristics of the CLAM steel manufactured using LPBF. Microcracks are another serious AM part defect that degrades the mechanical performance and fatigue life of the produced components. HIP has been shown to be an excellent post-processing method for eliminating internal microcracks (Tosi et al. 2022; Wang et al. 2021). However, studies indicated that only microcracks with a modest width (6 μm) were considerably reduced by the HIP process, while the number and length of the fractures do not influence the

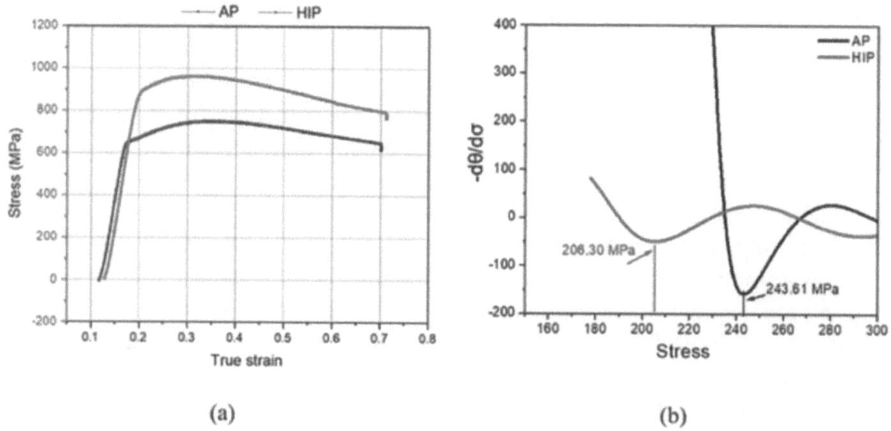

FIGURE 5.7 (a) Compressive true stress-strain curves of as-printed and HIP laser powder bed fusion specimens. (b) Critical stresses in as-printed and HIP samples during deformation (Sripada et al. 2022); reproduced after permission from Elsevier.

reduction of cracks in the HIP process. So far, research on HIP post-processing for AM components has shown some promising results. The HIP process is believed to decrease internal part defects and enhance fatigue properties significantly. However, current research on the HIP process is still in its early stages of investigation for AM. Further research is needed to better understand the HIP process for AM materials in order to obtain the most desirable properties.

5.3.4 Annealing: Solution and T6-Heat Treatments

Annealing is a process involving heat treatment applied to enhance the ductility and lessen the surface hardness of AM parts (Bryson 2015; Mahmood et al. 2022). In other words, heat treatment changes the materials' physical and sometimes chemical properties to make them more ductile and reduce their hardness. Normally, there are three stages in an annealing process, which are (a) recovery, (b) recrystallization, and (c) grain growth stage. In the first stage, the temperature of the material is raised, and the internal stresses are relieved. In the second stage, the material is heated above the recrystallization temperature but less than the melting point, resulting in new grain formation. In the third stage, the new grains are developed and regulated by permitting them to cool at some stated rate (Bryson 2015).

For AM-printed materials, solution and T-6 heat treatments are widely used as heat treatment process to improve their surface quality and increases corrosion resistance (Girelli et al. 2019; Mahmood et al. 2022). Solution heat treatment as a part of a consolidated process involving quenching and aging affects the mechanical properties of the printed parts. Many parameters are involved in solution heat treatment, such as time and temperature, which improve and affects the microstructure, surface roughness, and porosity. Ma et al. (2015) studied the effects of solution heat treatment on the mechanical properties of AA6082 alloys. The printed specimens were subjected to the solution heat treatment by a composite design and then processed by quenching and artificial aging for 3 h at 190 °C. According to their results, solution heat treatment time and temperature significantly affected the properties and fracture surfaces of the specimens after solution heat treatment compared to the controlled specimens. Increasing solution heat treatment time and temperature increased the ultimate tensile strength and decreased the ductility for the temperature range of 440 to 575 °C and time range of 1.7 to 58 min. Similarly, in the fracture surface of the specimens, some voids or dimple type sizes appeared after low solution heat treatment parameters, resulting in a ductile fracture mode.

Sing, Huang, and Yeong (2020) studied solution heat treatment on cobalt-28-chromium-6-molybdenum alloys printed by LPBF. They investigated the impact of solution heat treatment on the microstructure and mechanical characteristics of the printed specimens. A comparative study was done on their casted counterparts also. As-build specimens were anisotropic and presented an elevated microhardness in the *yz*-planes, while the solution heat treatment specimens showed similar values in both *yz*–and *xy*-planes. Compared with casted alloys, the as-build specimens demonstrated high mechanical characteristics with a focus on yield and ultimate tensile strengths as shown in Figure 5.8. After solution heat treatment, both strengths of the printed

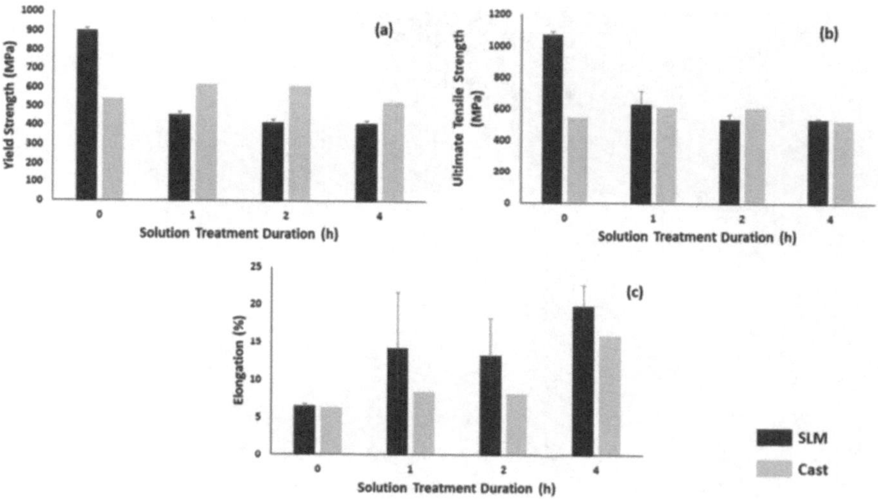

FIGURE 5.8 Effect of solution heat treatment on mechanical characteristics of Co-Cr-Mo fabricated by LPBF and casting process (a) yield strength, (b) ultimate tensile strength, and (c) elongation (Sing, Huang, and Yeong 2020); reproduced after permission from Elsevier.

specimens diminished. The strengths reduced gradually as the solution heat treatment time increased from 2 to 4 h. Due to solution heat treatment, the microstructure analysis results in the formation of carbides.

The influence of solution heat treatment time and temperature on the microstructure was studied, and investigated the oxidation performance of Inconel 718 (Calandri et al. 2018). Different trials were performed to check the capability of solution heat treatment, changing time and temperature, as shown in Figure 5.9. The excessive temperature of about 1200 °C lead to severe grain coarsening and poor mechanical properties. The optimal temperature of 1065 °C and dipping time of 2 h showed better efficiency in terms of maximal dissolution without any excessive growth rate. The extended oxidation run showed a steady parabolic rise in the mass when exposed to 850 °C in the open air. This mass grain per unit area was due to the development of a thick, protective, and steady oxide scale on the specimen surface.

Solution heat treatment decreases the surface roughness of the printed specimens. The surface roughness of AlSi10Mg and other parameters were investigated by Majeed et al. (2019). The average surface roughness of the printed specimens decreased up to 77% when SHT was done at 540 °C for 2 h; however, the surface roughness increased after aging at 155 °C for 12 h. Similarly, the influence of T6-heat treatment on the microstructure, hardness, and density of AlSi10Mg alloy was studied by Girelli et al. (2019). AlSi10Mg alloys were printed by direct metal laser sintering. Compared with gravity-casting specimens, the highest hardness and increase in porosity were achieved. Similarly, solution heat treatment decreased the ultimate tensile strength compared to as-built specimens.

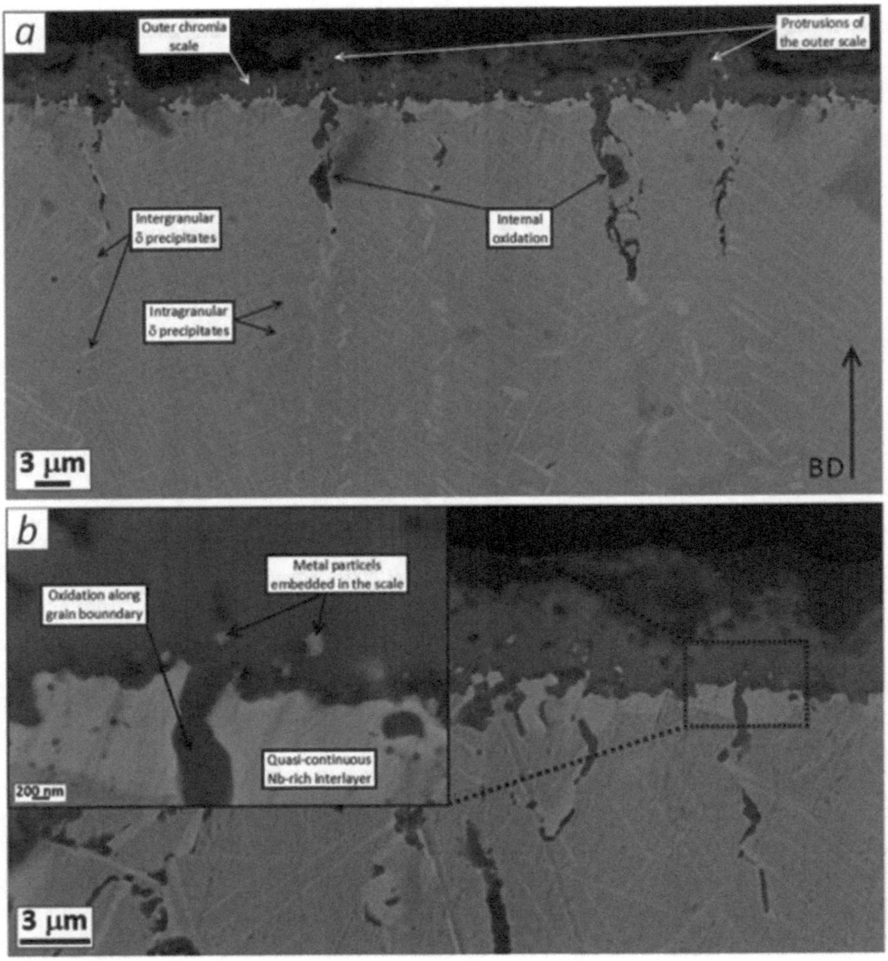

FIGURE 5.9 (a) Scanning electron micrographs of the oxide scale on solutioned LPBF Inconel-718 at 850 °C and (b) 908 h (Calandri et al. 2018); reproduced with permission from Wiley.

T6-heat treatment is a two-phase process applied to aluminum alloys manufactured by LPBF processes (Aboulkhair et al. 2016). Normally strength of up to 30% is increased as a result of this heat treatment. T6-heat treatment consists of three steps, which are (a) solution treatment, (b) quenching, and (c) aging. In the first step, a temperature of about 1200 °C is provided for the specimen. In quenching, the specimen is drastically cooled in a water bath. Artificial aging reheats the part a final time up to 600 °C . The microstructure and mechanical characteristics of AlSi10Mg alloy by LPBF were studied by Aboulkhair et al. (2016) and they investigated their effects on T6-heat treatment. Spatial variation was identified after T6-heat treatment due to phase evolution. Compared with die-cast parts, the printed specimen's microhardness was exceeded, and ultimate tensile strength of 333 MPa was achieved. Also, high compression strength was recorded for the T6-heat treatment parts compared with the as-built specimens.

Majeed et al. (2019) studied the effect of T6-heat treatment on the density and porosity of AlSi10Mg alloys manufactured by LPBF. According to their results, the best densification with minimum porosity was attained at the energy density of 144.89 J/mm^3. The densification was enhanced up to 99.87%, and low porosity was achieved due to T6-heat treatment. Similarly, the influence of T6-heat treatment temperature on the microstructure, hardness, and density of AlSi10Mg alloy was studied by Girelli et al. (2019). They studied the effects of T6-heat treatment on the corrosion behavior of AlSi10Mg alloys produced by LPBF. They revealed that after T6-heat treatment, the specimens achieved a high electrochemical corrosion resistance compared with the as-build specimens along with the lower mass. This enhancement was due to homogenization, and microstructure evaluation by the heat treatment resulted in the corrosion mechanism change. Effects of T6-heat treatment on mechanical properties, densification, hardness behavior, and oxidation behavior of AlSi10Mg alloys produced by LPBF were investigated by Yu and Wang (2018). The specimens undergo solid solution at 535 °C followed by artificial aging at 158 °C for 10 h. The hardness of T6-heat treatment samples decreased by about 20% compared to as-built specimens. This was attributed to forming of a fine-grained recrystallization microstructure during solid solution. Similarly, densification was rarely disturbed after T6-heat treatment. Overall, the solution heat treatment and T6-heat treatment processes substantially affect mechanical properties, porosity, density, microstructure, and surface roughness.

Table 5.1 provides a guideline to apply the specific post-processing technique based on the requirement. This guideline has been developed based on the trend identified in the literature. It is important to mention here that the hot-isostatic pressing has been applied to control the produced defects, while annealing has been used to improve the microstructure quality and corresponding mechanical properties.

TABLE 5.1
Guideline to Select a Post-Processing Technique for AM Parts

Process	Surface Roughness	Production Rate	Low Input Heat	Mechanical Properties	References
Laser shock peening	+/-	++	+++	+/-	(Rauch and Hascoet 2022)
Laser polishing	++	+/-	-	+/-	(Rauch and Hascoet 2022)
Hot-isostatic pressing	+/-	-	—	++	(Molaei, Fatemi, and Phan 2018)
Annealing	+/-	-	—	+++	(Molaei, Fatemi, and Phan 2018)

Note: "+" means positive and "-" means negative.

5.4 CONCLUSIONS

This chapter provides an overview of post-processes and their implications in AM. Various defects related to external and internal are produced in AM processes. To solve these issues, thermal post-processing techniques, such as laser shock peening, laser polishing, hot-isostatic pressing, and annealing techniques, such as solution heat treatment and T6-heat treatment are applied. These methods have demonstrated their potential to enhance mechanical properties, prevent residual stress generation, and improve AM items' surface polish. The following results have been deduced based on the current study:

- The laser shock peening process has been utilized for both thick and thin components. However, this process generates significant strain in the case of more fragile components. Local grain refinement occurs during laser shock peening, increasing hardness value. The intensity and wavelength of the laser beam also play a vital role in determining the surface regularities of a given material.
- Laser polishing controls surface roughness and hardness significantly. Laser polishing often increases the reflectivity of a surface by acting on its peaks. In addition, the number of passes regulates the surface characteristics of a specimen. Laser polishing has been shown to reduce surface roughness (= 95%) compared to the as-fabricated specimen.
- Commonly used annealing procedures, such as solution heat treatment and T6-heat treatment, remove pores, increase corrosion resistance, and enhance mechanical qualities. Using grain refinement and the compactness of deposited layers at extreme temperatures, these approaches reduce porosity by up to 99.99%.

REFERENCES

Aboulkhair, Nesma T., Ian Maskery, Chris Tuck, Ian Ashcroft, and Nicola M. Everitt. 2016. "The Microstructure and Mechanical Properties of Selectively Laser Melted AlSi10Mg: The Effect of a Conventional T6-like Heat Treatment." *Materials Science and Engineering: A* 667 (June): 139–46. https://doi.org/10.1016/J.MSEA.2016.04.092.

Aboulkhair, Nesma T., Marco Simonelli, Luke Parry, Ian Ashcroft, Christopher Tuck, and Richard Hague. 2019. "3D Printing of Aluminium Alloys: Additive Manufacturing of Aluminium Alloys Using Selective Laser Melting." *Progress in Materials Science* 106 (December): 100578. https://doi.org/10.1016/J.PMATSCI.2019.100578.

Atkinson, H V, and S Davies. 2000. "Fundamental Aspects of Hot Isostatic Pressing: An Overview." *Metallurgical and Materials Transactions A* 31 (12): 2981–3000.

Avilés, R., J. Albizuri, A. Lamikiz, E. Ukar, and A. Avilés. 2011. "Influence of Laser Polishing on the High Cycle Fatigue Strength of Medium Carbon AISI 1045 Steel." *International Journal of Fatigue* 33 (11): 1477–89. https://doi.org/10.1016/J.IJFATI GUE.2011.06.004.

Bryson, William E. 2015. "Heat Treatment." In *Heat Treatment*, edited by William E Bryson, I–2. Hanser. https://doi.org/https://doi.org/10.3139/9781569904862.fm.

Calandri, Michele, Diego Manfredi, Flaviana Calignano, Elisa Paola Ambrosio, Sara Biamino, Rocco Lupoi, and Daniele Ugues. 2018. "Solution Treatment Study of Inconel 718

Produced by SLM Additive Technique in View of the Oxidation Resistance." *Advanced Engineering Materials* 20 (11): 1800351. https://doi.org/https://doi.org/10.1002/adem.201800351.

Campbell, John. 2004. *Castings Practice: The Ten Rules of Castings*. Elsevier.

Cao, Sheng, Yichao Zou, Chao Voon Samuel Lim, and Xinhua Wu. 2021. "Review of Laser Powder Bed Fusion (LPBF) Fabricated Ti-6Al-4V: Process, Post-Process Treatment, Microstructure, and Property." *Light: Advanced Manufacturing* 2 (3): 313–32.

Carpenter, Kevin, and Ali Tabei. 2020. "On Residual Stress Development, Prevention, and Compensation in Metal Additive Manufacturing." *Materials* 13 (2). https://doi.org/10.3390/ma13020255.

Cegan, Tomas, Marek Pagac, Jan Jurica, Katerina Skotnicova, Jiri Hajnys, Lukas Horsak, Kamil Soucek, and Pavel Krpec. 2020. "Effect of Hot Isostatic Pressing on Porosity and Mechanical Properties of 316 l Stainless Steel Prepared by the Selective Laser Melting Method." *Materials* 13 (19): 4377.

Cerniglia, Donatella, and Nicola Montinaro. 2018. "Defect Detection in Additively Manufactured Components: Laser Ultrasound and Laser Thermography Comparison." *Procedia Structural Integrity* 8 (January): 154–62. https://doi.org/10.1016/J.PROSTR.2017.12.016.

Charalampous, Paschalis, Ioannis Kostavelis, and Dimitrios Tzovaras. 2020. "Non-Destructive Quality Control Methods in Additive Manufacturing: A Survey." *Rapid Prototyping Journal* 26 (4): 777–90. https://doi.org/10.1108/RPJ-08-2019-0224.

Chavez, Luis A., Paulina Ibave, Bethany Wilburn, David Alexander, Calvin Stewart, Ryan Wicker, and Yirong Lin. 2020. "The Influence of Printing Parameters, Post-Processing, and Testing Conditions on the Properties of Binder Jetting Additive Manufactured Functional Ceramics." *Ceramics 2020,* 3 (1): 65–77. https://doi.org/10.3390/CERAMICS3010008.

Chen, Chaoyue, Zhongming Ren, Yingchun Xie, Renzhong Huang, Min Liu, Xingchen Yan, Xinliang Xie, and Hanlin Liao. 2019. "Effect of Hot Isostatic Pressing (HIP) on Microstructure and Mechanical Properties of Ti6Ai4V Alloy Fabricated by Cold Spray Additive Manufacturing." In *Proceedings of the International Thermal Spray Conference*. Vol. 2019 (May).

Chen, Lan, Brodan Richter, Xinzhou Zhang, Kaila B. Bertsch, Dan J. Thoma, and Frank E. Pfefferkorn. 2021. "Effect of Laser Polishing on the Microstructure and Mechanical Properties of Stainless Steel 316L Fabricated by Laser Powder Bed Fusion." *Materials Science and Engineering: A* 802 (January): 140579. https://doi.org/10.1016/J.MSEA.2020.140579.

Chen, Yao, Xing Peng, Lingbao Kong, Guangxi Dong, Afaf Remani, and Richard Leach. 2021. "Defect Inspection Technologies for Additive Manufacturing." *International Journal of Extreme Manufacturing* 3 (2): 022002. https://doi.org/10.1088/2631-7990/ABE0D0.

Dai, Donghua, Dongdong Gu, Reinhart Poprawe, and Mujian Xia. 2017. "Influence of Additive Multilayer Feature on Thermodynamics, Stress and Microstructure Development during Laser 3D Printing of Aluminum-Based Material." *Science Bulletin* 62 (11): 779–87. https://doi.org/10.1016/J.SCIB.2017.05.007.

Dai, Lei, Ting Cheng, Chao Duan, Wei Zhao, Weipeng Zhang, Xuejun Zou, Joseph Aspler, and Yonghao Ni. 2019. "3D Printing Using Plant-Derived Cellulose and Its Derivatives: A Review." *Carbohydrate Polymers*. Elsevier Ltd. https://doi.org/10.1016/j.carbpol.2018.09.027.

Ding, K., and L. Ye. 2006. "General Introduction." *Laser Shock Peening*, January, 1–6. https://doi.org/10.1533/9781845691097.1.

Fairand, B. P., B. A. Wilcox, W. J. Gallagher, and D. N. Williams. 2003. "Laser Shock-induced Microstructural and Mechanical Property Changes in 7075 Aluminum." *Journal of Applied Physics* 43 (9): 3893. https://doi.org/10.1063/1.1661837.

Frazier, William E. 2014. "Metal Additive Manufacturing: A Review." *Journal of Materials Engineering and Performance* 23 (6): 1917–28. https://doi.org/10.1007/S11665-014-0958-Z/FIGURES/9.

Galy, Cassiopée, Emilie Le Guen, Eric Lacoste, and Corinne Arvieu. 2018. "Main Defects Observed in Aluminum Alloy Parts Produced by SLM: From Causes to Consequences." *Additive Manufacturing* 22 (August): 165–75. https://doi.org/10.1016/J.ADDMA.2018.05.005.

Girelli, Luca, Marialaura Tocci, Marcello Gelfi, and Annalisa Pola. 2019. "Study of Heat Treatment Parameters for Additively Manufactured AlSi10Mg in Comparison with Corresponding Cast Alloy." *Materials Science and Engineering: A* 739: 317–28. https://doi.org/https://doi.org/10.1016/j.msea.2018.10.026.

Grech, Iwan Salvu, J H Sullivan, Robert J Lancaster, J Plummer, and N P Lavery. 2022. "The Optimisation of Hot Isostatic Pressing Treatments for Enhanced Mechanical and Corrosion Performance of Stainless Steel 316L Produced by Laser Powder Bed Fusion." *Additive Manufacturing* 58: 103072.

Guo, Wei, Meng Hua, Peter Wai-Tat Tse, and Albert Chiu Kam Mok. 2011. "Process Parameters Selection for Laser Polishing DF2 (AISI O1) by Nd:YAG Pulsed Laser Using Orthogonal Design." *The International Journal of Advanced Manufacturing Technology 2011 59:9* 59 (9): 1009–23. https://doi.org/10.1007/S00170-011-3558-1.

Hackel, Lloyd, Jon R. Rankin, Alexander Rubenchik, Wayne E. King, and Manyalibo Matthews. 2018. "Laser Peening: A Tool for Additive Manufacturing Post-Processing." *Additive Manufacturing* 24 (December): 67–75. https://doi.org/10.1016/J.ADDMA.2018.09.013.

Herzog, Dirk, Vanessa Seyda, Eric Wycisk, and Claus Emmelmann. 2016. "Additive Manufacturing of Metals." *Acta Materialia* 117 (September): 371–92. https://doi.org/10.1016/J.ACTAMAT.2016.07.019.

Jinoop, A. N., S. Kanmani Subbu, C. P. Paul, and I. A. Palani. 2019. "Post-Processing of Laser Additive Manufactured Inconel 718 Using Laser Shock Peening." *International Journal of Precision Engineering and Manufacturing* 20 (9): 1621–28. https://doi.org/10.1007/S12541-019-00147-4.

Kempen, K., L. Thijs, E. Yasa, M. Badrossamay, W. Verheecke, and J.-P. Kruth. 2011. "Process Optimization and Microstructural Analysis for Selective Laser Melting of AlSi10Mg," August. https://doi.org/10.26153/TSW/15310.

Kimura, Takahiro, and Takayuki Nakamoto. 2016. "Microstructures and Mechanical Properties of A356 (AlSi7Mg0.3) Aluminum Alloy Fabricated by Selective Laser Melting." *Materials & Design* 89 (January): 1294–1301. https://doi.org/10.1016/J.MATDES.2015.10.065.

Ko, Hyunwoong, Seung Ki Moon, and Jihong Hwang. 2015. "Design for Additive Manufacturing in Customized Products." *International Journal of Precision Engineering and Manufacturing* 16 (11): 2369–75. https://doi.org/10.1007/S12541-015-0305-9.

Kong, Lingbao, Xing Peng, Yao Chen, Ping Wang, and Min Xu. 2020. "Multi-Sensor Measurement and Data Fusion Technology for Manufacturing Process Monitoring: A Literature Review." *International Journal of Extreme Manufacturing* 2 (2): 022001. https://doi.org/10.1088/2631-7990/AB7AE6.

Kou, Sindo. 2015. "A Criterion for Cracking during Solidification." *Acta Materialia* 88 (April): 366–74. https://doi.org/10.1016/J.ACTAMAT.2015.01.034.

Kruth, Jean-Pierre, Mohsen Badrossamay, Evren Yasa, Jan Deckers, Lore Thijs, and Jan Van Humbeeck. 2010. "Part and Material Properties in Selective Laser Melting of Metals." *Proceedings of the 16th International Symposium on Electromachining (ISEM XVI)* 9: 3–14. https://lirias.kuleuven.be/66146.

Lamikiz, A., J. A. Sánchez, L. N. López de Lacalle, and J. L. Arana. 2007. "Laser Polishing of Parts Built up by Selective Laser Sintering." *International Journal of Machine Tools and Manufacture* 47 (12–13): 2040–50. https://doi.org/10.1016/J.IJMACHTOOLS.2007.01.013.

Lan, Liang, Ruyi Xin, Xinyuan Jin, Shuang Gao, Bo He, Yonghua Rong, and Na Min. 2020. "Effects of Laser Shock Peening on Microstructure and Properties of Ti–6Al–4V Titanium Alloy Fabricated via Selective Laser Melting." *Materials* 13 (15): 3261. https://doi.org/10.3390/MA13153261.

Lavery, N P, John Cherry, Shahid Mehmood, Helen Davies, B Girling, Elizabeth Sackett, S G R Brown, and Johann Sienz. 2017. "Effects of Hot Isostatic Pressing on the Elastic Modulus and Tensile Properties of 316L Parts Made by Powder Bed Laser Fusion." *Materials Science and Engineering: A* 693: 186–213.

Lee, Seungjong, Zabihollah Ahmadi, Jonathan W. Pegues, Masoud Mahjouri-Samani, and Nima Shamsaei. 2021. "Laser Polishing for Improving Fatigue Performance of Additive Manufactured Ti-6Al-4V Parts." *Optics & Laser Technology* 134 (February): 106639. https://doi.org/10.1016/J.OPTLASTEC.2020.106639.

Li, Wei, Shuai Li, Jie Liu, Ang Zhang, Yan Zhou, Qingsong Wei, Chunze Yan, and Yusheng Shi. 2016. "Effect of Heat Treatment on AlSi10Mg Alloy Fabricated by Selective Laser Melting: Microstructure Evolution, Mechanical Properties and Fracture Mechanism." *Materials Science and Engineering: A* 663 (April): 116–25. https://doi.org/10.1016/J.MSEA.2016.03.088.

"Library." 2003. *MRS Bulletin* 28 (9): 674–75. https://doi.org/10.1557/mrs2003.197.

Liu, Xiaohui, Yunzhong Liu, Zhiguang Zhou, Huan Zhong, and Qiangkun Zhan. 2022. "A Combination Strategy for Additive Manufacturing of AA2024 High-Strength Aluminium Alloys Fabricated by Laser Powder Bed Fusion: Role of Hot Isostatic Pressing." *Materials Science and Engineering: A* 850: 143597.

Liverani, Erica, Adrian H A Lutey, Alessandro Ascari, and Alessandro Fortunato. 2020. "The Effects of Hot Isostatic Pressing (HIP) and Solubilization Heat Treatment on the Density, Mechanical Properties, and Microstructure of Austenitic Stainless Steel Parts Produced by Selective Laser Melting (SLM)." *The International Journal of Advanced Manufacturing Technology* 107 (1): 109–22.

Louvis, Eleftherios, Peter Fox, and Christopher J. Sutcliffe. 2011. "Selective Laser Melting of Aluminium Components." *Journal of Materials Processing Technology* 211 (2): 275–84. https://doi.org/10.1016/J.JMATPROTEC.2010.09.019.

Ma, C. P., Y. C. Guan, and W. Zhou. 2017. "Laser Polishing of Additive Manufactured Ti Alloys." *Optics and Lasers in Engineering* 93 (June): 171–77. https://doi.org/10.1016/J.OPTLASENG.2017.02.005.

Ma, Wenyu, Baoyu Wang, Lei Yang, Xuefeng Tang, Wenchao Xiao, and Jing Zhou. 2015. "Influence of Solution Heat Treatment on Mechanical Response and Fracture Behaviour of Aluminium Alloy Sheets: An Experimental Study." *Materials & Design* 88: 1119–26. https://doi.org/https://doi.org/10.1016/j.matdes.2015.09.044.

Mahmood, Muhammad Arif, Andrei C. Popescu, Claudiu Liviu Hapenciuc, Carmen Ristoscu, Anita Ioana Visan, Mihai Oane, and Ion N. Mihailescu. 2020. "Estimation of Clad Geometry and Corresponding Residual Stress Distribution in Laser Melting Deposition: Analytical Modeling and Experimental Correlations." *The International*

Journal of Advanced Manufacturing Technology 111 (September): 77–91. https://doi.org/10.1007/s00170-020-06047-6.

Mahmood, Muhammad Arif, Diana Chioibasu, Asif Ur Rehman, Sabin Mihai, and Andrei C Popescu. 2022. “Post-Processing Techniques to Enhance the Quality of Metallic Parts Produced by Additive Manufacturing.” *Metals* 12 (1): 77. https://doi.org/10.3390/MET12010077.

Mai, T. A., and G. C. Lim. 2004. “Micromelting and Its Effects on Surface Topography and Properties in Laser Polishing of Stainless Steel.” *Journal of Laser Applications* 16 (4): 221. https://doi.org/10.2351/1.1809637.

Majeed, Arfan, Altaf Ahmed, Abdus Salam, and Muhammad Zakir Sheikh. 2019. “Surface Quality Improvement by Parameters Analysis, Optimization and Heat Treatment of AlSi10Mg Parts Manufactured by SLM Additive Manufacturing.” *International Journal of Lightweight Materials and Manufacture* 2 (4): 288–95. https://doi.org/https://doi.org/10.1016/j.ijlmm.2019.08.001.

Molaei, Reza, Ali Fatemi, and Nam Phan. 2018. “Significance of Hot Isostatic Pressing (HIP) on Multiaxial Deformation and Fatigue Behaviors of Additive Manufactured Ti-6Al-4V Including Build Orientation and Surface Roughness Effects.” *International Journal of Fatigue* 117 (December): 352–70. https://doi.org/10.1016/J.IJFATIGUE.2018.07.035.

Morgan, Victor T. 2014. “The Effect of Porosity on Some of The Physical Properties of Powder-Metallurgy Components.” 6 (12): 72–86. https://doi.org/10.1179/POM.1963.6.12.006.

Olakanmi, E. O. 2013. “Selective Laser Sintering/Melting (SLS/SLM) of Pure Al, Al–Mg, and Al–Si Powders: Effect of Processing Conditions and Powder Properties.” *Journal of Materials Processing Technology* 213 (8): 1387–1405. https://doi.org/10.1016/J.JMATPROTEC.2013.03.009.

Pandey, Pulak M., N. Venkata Reddy, and Sanjay G. Dhande. 2003. “Improvement of Surface Finish by Staircase Machining in Fused Deposition Modeling.” *Journal of Materials Processing Technology* 132 (1–3): 323–31. https://doi.org/10.1016/S0924-0136(02)00953-6.

Plessis, A. Du, and E. Macdonald. 2020. “Hot Isostatic Pressing in Metal Additive Manufacturing: X-Ray Tomography Reveals Details of Pore Closure, Addit. Manuf. 34 (2020) 101191.”

Puichaud, Anne-Helene, Camille Flament, Aziz Chniouel, Fernando Lomello, Elodie Rouesne, Pierre-François Giroux, Hicham Maskrot, Frederic Schuster, and Jean-Luc Béchade. 2019. “Microstructure and Mechanical Properties Relationship of Additively Manufactured 316L Stainless Steel by Selective Laser Melting.” *EPJ Nuclear Sciences & Technologies* 5: 23.

Pyka, Grzegorz, Greet Kerckhofs, Ioannis Papantoniou, Mathew Speirs, Jan Schrooten, and Martine Wevers. 2013. “Surface Roughness and Morphology Customization of Additive Manufactured Open Porous Ti6Al4V Structures.” *Materials* 6 (10): 4737–57. https://doi.org/10.3390/MA6104737.

Raja, Kumar, Manoj Nathan, Tejas Patil Balram, and C. D. Naiju. 2018. “Study of Surface Integrity and Effect of Laser Peening on Maraging Steel Produced by Lasercusing Technique.” *SAE Technical Papers* 2018 (July). https://doi.org/10.4271/2018-28-0094.

Rappaz, M., J. M. Drezet, and M. Gremaud. 1999a. “A New Hot-Tearing Criterion.” *Metallurgical and Materials Transactions A* 30 (2): 449–55. https://doi.org/10.1007/S11661-999-0334-Z.

Rappaz, M, J.-M Drezet, and M Gremaud. 1999b. “A New Hot-Tearing Criterion.” *Metallurgical and Materials Transactions A* 450: 449.

Rauch, Matthieu, and Jean Yves Hascoet. 2022. "A Comparison of Post-Processing Techniques for Additive Manufacturing Components." *Procedia CIRP* 108 (C): 442–47. https://doi.org/10.1016/j.procir.2022.03.069.

Rosa, Benoit, Pascal Mognol, and Jean-yves Hascoët. 2015. "Laser Polishing of Additive Laser Manufacturing Surfaces." *Journal of Laser Applications* 27 (S2): S29102. https://doi.org/10.2351/1.4906385.

Salimianrizi, A., E. Foroozmehr, M. Badrossamay, and H. Farrokhpour. 2016. "Effect of Laser Shock Peening on Surface Properties and Residual Stress of Al6061-T6." *Optics and Lasers in Engineering* 77 (February): 112–17. https://doi.org/10.1016/J.OPTLASENG.2015.08.001.

Shiyas, K A, and R Ramanujam. 2021. "A Review on Post Processing Techniques of Additively Manufactured Metal Parts for Improving the Material Properties." *Materials Today: Proceedings* 46: 1429–36.

Sidhu, Kuldeep Singh, Yachao Wang, Jing Shi, Vijay K. Vasudevan, and Seetha Ramaiah Mannava. 2019. "Effect of Post Laser Shock Peening on Microstructure and Mechanical Properties of Inconel 718 by Selective Laser Melting." *ASME 2019 14th International Manufacturing Science and Engineering Conference, MSEC 2019* 2 (November). https://doi.org/10.1115/MSEC2019-2893.

Sing, Swee Leong, Sheng Huang, and Wai Yee Yeong. 2020. "Effect of Solution Heat Treatment on Microstructure and Mechanical Properties of Laser Powder Bed Fusion Produced Cobalt-28chromium-6molybdenum." *Materials Science and Engineering: A* 769: 138511. https://doi.org/https://doi.org/10.1016/j.msea.2019.138511.

Sinico, Mirko, Suraj Dinkar Jadhav, Ann Witvrouw, Kim Vanmeensel, and Wim Dewulf. 2021. "A Micro-Computed Tomography Comparison of the Porosity in Additively Fabricated CuCr1 Alloy Parts Using Virgin and Surface-Modified Powders." *Materials* 14 (8). https://doi.org/10.3390/MA14081995.

Sripada, Jagannadh, Yuan Tian, Kanwal Chadha, Gobinda Saha, Mohammad Jahazi, John Spray, and Clodualdo Aranas Jr. 2022. "Effect of Hot Isostatic Pressing on Microstructural and Micromechanical Properties of Additively Manufactured 17–4PH Steel." *Materials Characterization* 192: 112174.

Sun, Z., X. P. Tan, M. Descoins, D. Mangelinck, S. B. Tor, and C. S. Lim. 2019. "Revealing Hot Tearing Mechanism for an Additively Manufactured High-Entropy Alloy via Selective Laser Melting." *Scripta Materialia* 168 (July): 129–33. https://doi.org/10.1016/J.SCRIPTAMAT.2019.04.036.

Sundar, R., P. Ganesh, Ram Kishor Gupta, G. Ragvendra, B. K. Pant, Vivekanand Kain, K. Ranganathan, Rakesh Kaul, and K. S. Bindra. 2019. "Laser Shock Peening and Its Applications: A Review." *Lasers in Manufacturing and Materials Processing* 6 (4): 424–63. https://doi.org/10.1007/S40516-019-00098-8/FIGURES/7.

Taheri, Hossein, Mohammad Rashid Bin Mohammad Shoaib, Lucas W. Koester, Timothy A. Bigelow, Peter C. Collins, and Leonard J. Bond. 2017. "Powder-Based Additive Manufacturing–a Review of Types of Defects, Generation Mechanisms, Detection, Property Evaluation and Metrology." *International Journal of Additive and Subtractive Materials Manufacturing* 1 (2): 172. https://doi.org/10.1504/IJASMM.2017.088204.

Tan, Hua, Yanbo Fang, Chongliang Zhong, Zihao Yuan, Wei Fan, Zuo Li, Jing Chen, and Xin Lin. 2020. "Investigation of Heating Behavior of Laser Beam on Powder Stream in Directed Energy Deposition." *Surface and Coatings Technology* 397 (September): 126061. https://doi.org/10.1016/J.SURFCOAT.2020.126061.

Thijs, Lore, Karolien Kempen, Jean Pierre Kruth, and Jan Van Humbeeck. 2013. "Fine-Structured Aluminium Products with Controllable Texture by Selective Laser Melting of Pre-Alloyed AlSi10Mg Powder." *Acta Materialia* 61 (5): 1809–19. https://doi.org/10.1016/J.ACTAMAT.2012.11.052.

Thompson, A., I. Maskery, and R. K. Leach. 2016. "X-Ray Computed Tomography for Additive Manufacturing: A Review." *Measurement Science and Technology* 27 (7). https://doi.org/10.1088/0957-0233/27/7/072001.

Tomus, Dacian, Paul A. Rometsch, Martin Heilmaier, and Xinhua Wu. 2017. "Effect of Minor Alloying Elements on Crack-Formation Characteristics of Hastelloy-X Manufactured by Selective Laser Melting." *Additive Manufacturing* 16 (August): 65–72. https://doi.org/10.1016/J.ADDMA.2017.05.006.

Tosi, Riccardo, Chu Lun Alex Leung, Xipeng Tan, Emmanuel Muzangaza, and Moataz M Attallah. 2022. "Revealing the Microstructural Evolution of Electron Beam Powder Bed Fusion and Hot Isostatic Pressing Ti-6Al-4V in-Situ Shelling Samples Using X-Ray Computed Tomography." *Additive Manufacturing* 57: 102962.

Velu, Rajkumar, Arun V. Kumar, A. S. S. Balan, and Jyoti Mazumder. 2021. "Laser Aided Metal Additive Manufacturing and Postprocessing: A Comprehensive Review." *Additive Manufacturing*, January, 427–56. https://doi.org/10.1016/B978-0-12-818411-0.00023-9.

Wang, Hui, Liu Chen, Bogdan Dovgyy, Wenyong Xu, Aixue Sha, Xingwu Li, Huiping Tang, Yong Liu, Hong Wu, and Minh-Son Pham. 2021. "Micro-Cracking, Microstructure and Mechanical Properties of Hastelloy-X Alloy Printed by Laser Powder Bed Fusion: As-Built, Annealed and Hot-Isostatic Pressed." *Additive Manufacturing* 39: 101853.

Wang, Qinghua, Justin D. Morrow, Chao Ma, Neil A. Duffie, and Frank E. Pfefferkorn. 2015. "Surface Prediction Model for Thermocapillary Regime Pulsed Laser Micro Polishing of Metals." *Journal of Manufacturing Processes* 20 (October): 340–48. https://doi.org/10.1016/J.JMAPRO.2015.05.005.

Yu, Xianglong, and Lianfeng Wang. 2018. "T6 Heat-Treated AlSi10Mg Alloys Additive-Manufactured by Selective Laser Melting." *Procedia Manufacturing* 15: 1701–7. https://doi.org/https://doi.org/10.1016/j.promfg.2018.07.265.

Zerbst, Uwe, and Mauro Madia. 2021. *Structural Integrity II: Fatigue Properties. Fundamentals of Laser Powder Bed Fusion of Metals*. Elsevier. https://doi.org/10.1016/B978-0-12-824090-8.00015-9.

Zhai, Yutao, Bo Huang, Xiaodong Mao, and Mingjie Zheng. 2019. "Effect of Hot Isostatic Pressing on Microstructure and Mechanical Properties of CLAM Steel Produced by Selective Laser Melting." *Journal of Nuclear Materials* 515: 111–21.

Zhang, Jiong, Alvin You Xiang Toh, Hao Wang, Wen Feng Lu, and Jerry Ying Hsi Fuh. 2018. "Vibration-Assisted Conformal Polishing of Additively Manufactured Structured Surface:" 233 (12): 4154–64. https://doi.org/10.1177/0954406218811359.

Zhou, Jing, Conghao Liao, Hong Shen, and Xiaohong Ding. 2019. "Surface and Property Characterization of Laser Polished Ti6Al4V." *Surface and Coatings Technology* 380 (December): 125016. https://doi.org/10.1016/J.SURFCOAT.2019.125016.

6 Chemical-Based Post-Processing in Additive Manufacturing Processes

Iliyas Melnikov[1,3], Baltej Singh Rupal[2], and Hamid Rajani[1]
[1]Southern Alberta Institute of Technology (SAIT), Canada
[2]University of Alberta, Canada
[3]University College Leuven-Limburg (UCLL), Belgium

CONTENTS

6.1 INTRODUCTION

These days, it is common knowledge that additive manufacturing (AM) is a process that utilizes computer-aided-design (CAD) models to manufacture a three-dimensional (3D) part. This is usually achieved by material addition in a layer-by-layer style. The 3D model is generated using CAD software. However, it is also a possibility to use tools, such as 3D scanners to generate a 3D model of an already existing object. The 3D model is then converted to standard tessellation language (STL) file format, which is easier to process for the slicing software. The triangulated STL file is then imported to a slicing software to generate a layer-by-layer toolpath and for assigning manufacturing process parameters to each layer. The toolpath or the G-code is then fed to the AM machine to start the manufacturing process [1].

There are many types of AM processes available commercially and in the research stage. These AM processes already have a vast number of applications such as in the biomedical industry, automotive industry, prototype testing, and so on. However, for the sake of this chapter, the focus will be on the two most typically used AM

DOI: 10.1201/9781003276111-6

processes for manufacturing polymer parts, i.e., fused filament fabrication (FFF) and vat photopolymerization (VPP). The subsections below will give a brief overview of the abovementioned polymer-based AM processes with a focus on materials, process parameters, and output part properties [2, 3].

6.1.1 Fused Filament Fabrication

FFF uses a polymer filament and a material extruder to manufacture the part. During the FFF manufacturing process, a polymer filament feedstock is fed to a heated nozzle, melted, and then selectively deposited in a layer-by-layer fashion as seen in Figure 6.1. The type of feedstock directly links to the process parameters as well as the output properties of the manufactured part. Because of this, it has become standard procedure to do the material selection during the part designing process. Several polymer filaments can be processed using the FFF method, some of the most widely used are discussed in Table 6.1 along with their material properties, significant characteristics, and application areas.

Depending on the filament material, systematic experimental investigation of the process parameters should also be considered to guarantee an optimal manufactured part in terms of part properties [11]. Some of the important process parameters along with their effect on the part properties are shown in Table 6.2.

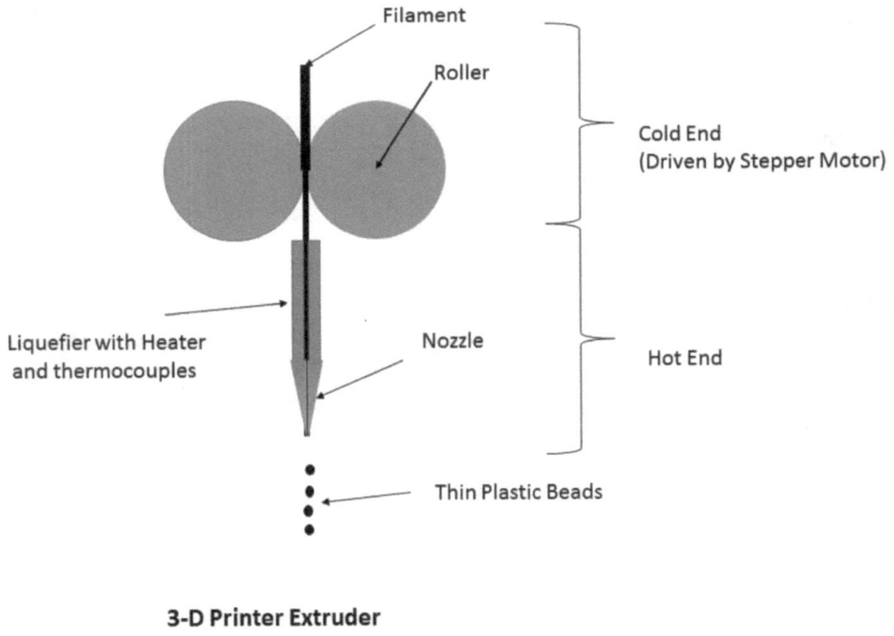

FIGURE 6.1 Schematic of FFF process [4].

TABLE 6.1
Different Feedstock Polymers for the FFF Process [5–10]

Polymer	Printing Temp. (°C)	Ultimate Tensile Strength (MPa)	Characteristics	Applications
PLA	190–230	25–50	• Odorless • Biodegradable • UV resistance • Low humidity resistance	• Medical implants • Prototyping
ABS	210–235	22–74	• Great abrasion resistance • Glue able with acetone • UV sensitive • Odor while printing	• Automotive industry • Electrical industry
PETG	230–250	20–69	• High resistance to humidity and chemicals • Contact with food is safe for human consumption • Good resistance to abrasion • Recyclable	• Food and beverage containers • Medical and pharmaceutical industry
PC	260–310	28–75	• Sterilizable • UV sensitive	• Security glass • Automotive industry
Nylon	240–270	50–90	• High strength • Good chemical resistance • Absorbs moisture	• Fishing industry • Adventure travel industry
TPU	220–250	28–96	• Resistant to oil and grease, • Good abrasion resistance • Hard to glue	• Automotive industry • Sport industry • Power tools

TABLE 6.2
Different Process Parameters for FFF and Their Effect on Output Properties [11]

Process Parameter	Effect on Output Properties
Layer thickness	Impacts: visual appearance, dimensional accuracy, surface quality, and build time.
Build orientation	Determines if support material is needed. Can increase or decrease tensile strength.
Extrusion temperature	Determines how well filament can be extruded as well as the adherence between layers.
Infill density	Impacts: density of part, weight, material consumption, build time, stability and durability, compressive strength.
Infill pattern	Impacts structure and mechanical properties.
Number of shells	Impacts part stability and sturdiness, material consumption, build time, aesthetics, and structural properties.
Raster width	Impacts printed layer thickness and build time.
Print speed	Affects the build time and quality of the printed part.

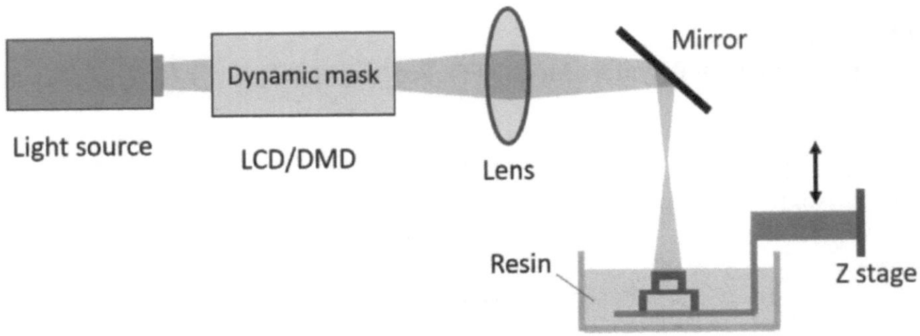

FIGURE 6.2 A generic schematic of the VPP process [16].

6.1.2 Vat Photopolymerization

VPP a type of additive manufacturing process that utilizes the hardening properties of a resin-based material. There are two types of VPP, digital light processing (DLP) and stereolithography (SLA). Both use liquid photopolymer resin, which is then solidified by exposure to a UV light source. The light source used in DLP is a digital light projector, this flashes the layers of the part sequentially. Since the light source projects, an image of the layer, all points in the layer will cure simultaneously. In contrast to SLA, which uses a laser to cure one point (or pixel) at a time, DLP is a much faster process [12]. Once the layer has completed its curing process, the build platform is moved by one layer height in the *z*-axis direction. After which the next layer is cured on the previous layer, and the process is repeated until a complete part is manufactured, as seen in Figure 6.2.

The liquid photopolymer resin in VPP drastically differs from the feedstock used in FFF. Instead of one polymer source for each filament type, photopolymer resins contain multiple monomers, as well as other ingredients that are necessary to initiate photopolymerization [13–14]. In theory, the combinations of monomers can be endless, however, for specific applications, certain types of resins have been developed, and some of them are discussed briefly in Table 6.3. As is the case with FFF, process parameters for VPP also play a significant role in determining the part properties [16]. For VPP, the significant process parameters and their effect on the output part properties are depicted in Table 6.4.

Optimizing the process parameters becomes paramount when it comes to assuring the quality of the AM part properties [17]. The most important output properties for AM parts are outlined below:

- mechanical properties;
- dimensional accuracy;
- surface topology.

Mechanical properties of a printed part such as hardness can be improved by physical post-processing methods like annealing. In the case of improving surface

TABLE 6.3
Different Resin Types and Their Major Properties [12, 13]

Resin Type	Properties	Example
Standard resin	Smooth surface quality, affordable, resistance to impact and breaking is low, available in the largest range of colors, highlights details and features well, not intended for functional parts.	Formlabs White Resin, Elegoo Mars Standard Resin
Clear resin	Transparent parts possible, water-resistant, great surface finish, susceptible to UV degradation over time.	Formlabs Clear Resin, Anycubic Clear 3D Resin
Tough resin	High resistance to shattering and impact, very durable and sturdy, removal of print is difficult, best utilized for strong functional parts.	Formlabs Tough 2000 Resin, eSUN Tough High Impact Resin
Flexible resin	Great flexibility, good impact resistance, hard to print, required use of support structures,	Formlabs Elastic 50A, Siraya Tech Flexible Tenacious Resin
Water-washable resin	Easiest to post-process, good strength, fast to cure, great detail, possible to dye, can absorb water.	Elegoo Water Washable Ceramic Gray Resin
Dental resin	Good compressive strength, resistance to abrasion with enamel, decent resistance to fractures.	Formlabs Permanent Crown resin, EPAX 3D Printer Dental Resin
Ceramic-filled resin	High stiffness and toughness, high resistance to heat, brittle, and lesser impact strength.	Formlabs Ceramic Resin, iFUN Ceramics Resin

TABLE 6.4
Different Process Parameters for VPP and Their Effect on Output Properties [16]

Process Parameter	Effect on Output Properties
Layer thickness	Impacts: visual appearance, dimensional accuracy, surface quality, and build time.
Build orientation	Determines if support material is needed. Can increase or decrease tensile strength.
Exposure time	Determines how well the adhesion happens between different layers.
Print Speed	Affects: build time, quality of the printed part.

topology, it can be achieved by different post-processing methods. Mostly these post-processing methods are physical methods, for example, surface sanding and remelting to reduce surface roughness. However, these methods do come at the cost of dimensional accuracy. In comparison to chemical post-processing, physical/other post-processing methods are well incorporated into industrial workflows [18].

Although chemical post-processing methods often produce superior surface topology and are relatively easier to implement than their physical counterparts. The reason behind this is the lack of good resources and knowledge to understand and implement the chemical post-processing methods for polymer-based AM processes. Therefore, this chapter will introduce and explain different chemical post-processing methods for polymer-based AM processes.

6.2 CHEMICAL POST-PROCESSING

As discussed before, additive manufacturing has many advantages when compared to conventional subtractive manufacturing, such as shape complexity, no tooling requirements, and so on. However, AM still has some drawbacks, such as the staircase effect (refer to Figure 6.3) which leads to a lack of dimensional accuracy and surface quality issues. To improve these properties, different post-processing methods can be implemented [18]. The post-processing stage starts after the AM part has finished manufacturing and can be divided into further stages. The first stage of the post-processing includes the removal of the part from the base plate and support structures, as well as generic cleaning. These simple post-processing steps can suffice for some applications that do not require good surface qualities. The second stage often includes post-processing methods that aim to improve the surface finish of AM parts [19]. Since AM parts possess intricate geometries and organic shapes, it becomes difficult to use physical post-processing methods to increase the surface quality. A solution to this could be chemical post-processing methods. Chemical post-processing methods make use of chemicals to treat the manufactured part and to achieve the desired surface properties. Once set up, the method requires relatively low human labor, is easy to use, and has the potential to be fully automated [21]. There are different types of chemical post-processing methods that have been and can be implemented successfully for AM parts. Some of them are as follows:

1. chemical vapor smoothing;
2. electro/electroless plating;
3. polymer coatings.

FIGURE 6.3 Visualization of the staircase effect on AM parts (physical part vs. CAD model).

6.2.1 Chemical Vapor Smoothing

It is still debated whether FFF parts are suitable for precision engineering due to the rough textures of the part. These rough textures are mostly caused by the staircase effect. Due to the layer-by-layer fabrication technique, the stair-casing effect leads to deviation in the FFF part as compared to the input CAD model [21], as illustrated in Figure 6.3.

The use of chemical vapor smoothing can effectively increase the surface finish of thermoplastic FFF parts made of polylactic acid (PLA) or ABS (acrylonitrile butadiene styrene) [22]. To produce a polished end-product from this post-processing technique, the polymer needs to be soluble in the solvent. Otherwise, surface polishing will not occur. Thus, a good understanding of polymer dissolution behavior is needed to optimize and apply this post-processing method.

In contrast to non-polymeric materials, polymers do not dissolve instantly. The rate of dissolution in polymers is mainly controlled by two transport processes, namely chain disentanglement and solvent diffusion. In the presence of a compatible solvent, an amorphous polymer will start to swell. This is due to the solvent acting as a plasticizer and diffusing through the surface of the polymer part, resulting in a gel-like outer layer. Figure 6.4 illustrates how this results in the formation of two interfaces, one between the solvent and gel layer, and one between the gel and solid layer. However, this is not always true, as a few polymers exhibit cracking instead of swelling [23].

The vapor smoothing post-processing technique consists of five individual steps, as illustrated in Figure 6.5 [27].

In the beginning, the untreated FFF part exhibits poor surface properties, such as the staircase effect and high surface roughness (step 1). The part is then placed in a specialized chamber where hot chemical vapors will interact with its surface (step 2). In step 3, the material on the surface forms a molten/gel-like layer (step 3). In the fourth step, the dissolved gel-like material will start flowing into the gaps between the

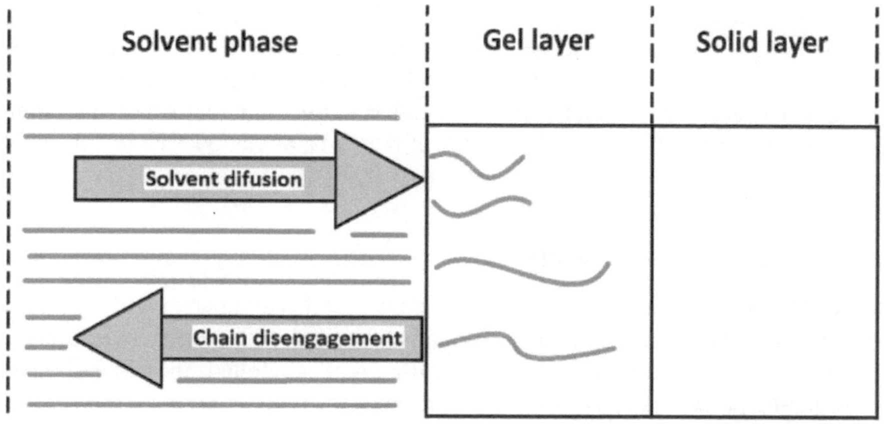

FIGURE 6.4 One-dimensional view of solvent diffusion and polymer chain disengagement.

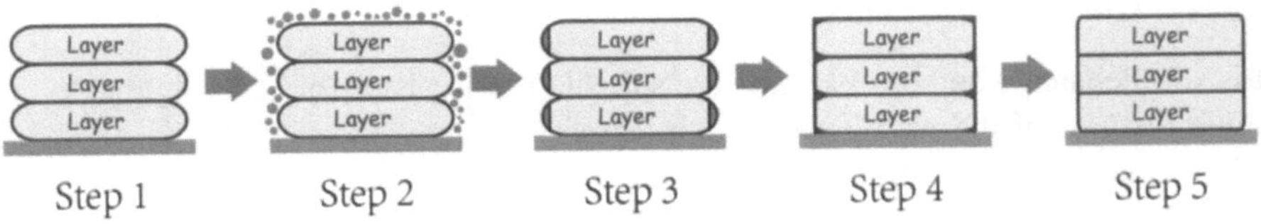

FIGURE 6.5 Schematic of the different steps in chemical vapor smoothing [27].

TABLE 6.5
Surface Roughness of FFF Parts, Measured after Each Cycle [28]

	Average Surface Roughness (Ra)				
Test Number	Before Treatment (μm)	After First Cycle (μm)	After Second Cycle (μm)	After Third Cycle (μm)	Total Improvement (%)
1	9.06	0.83	0.47	0.39	94.8
2	9.05	0.64	0.42	0.31	95.4
3	9.05	0.51	0.31	0.21	96.5

layers and fill them up. This happens because the material experiences relatively high surface tension by holding up the volume of the staircase structure. Due to this phenomenon, the molten gel-like surface material will seek to reduce the surface tension by decreasing its contact surface with the air around it. And thus, creating a flat surface (step 4). Finally, the vapor smoothing process is completed, once the surface of the part is solidified (step 5) [27, 28].

Multiple studies conclude that a decrease in surface roughness is found when chemical vapor smoothing is applied to post-process FFF parts. One of the studies used a hip implant made out of ABS, the implant was produced with FFF, and the results of the study can be seen in Table 6.5 [28]. Another study found a similar result but instead of using a geometrically complicated part, the authors opted for simpler parts like cubes, cylinders, and hemispheres made from ABS. The simpler geometry also enabled the calculation and data collection of the dimensional deviation of the parts. The study concluded that it is possible to use chemical vapor smoothing without affecting the dimensional accuracy of the parts, as can be seen in Tables 6.6 and 6.7 [26].

6.2.2 Electro/Electroless Plating

The electroplating process uses an anode made of metal coating and a cathode that is the conductive substrate. Both the anode and cathode are submerged in an aqueous electrolyte solution after which an electrical current is applied, thus initiating the electroplating process [29] as shown in Figure 6.6.

Electroplating makes use of electrochemical methods to produce a metallic film on metallic surfaces. In contrast to other methods, this method is already well explored

TABLE 6.6
Surface Roughness of FFF Parts, Measured before and after Treatment [26]

Test Number	Form	Average Surface Roughness: Before Treatment (μm)	Average Surface Roughness: After Treatment (μm)	Treatment Time (s)	Improvement (%)
1	Cube	8.45	0.39	10	95.4
2	Cube	9.18	019	15	97.9
3	Cube	8.83	0.07	20	99.2
4	Cylinder	8.60	0.18	15	97.9
5	Cylinder	8.59	0.07	20	99.1
6	Cylinder	8.76	0.24	10	97.2
7	Hemisphere	8.60	0.08	20	99.1
8	Hemisphere	8.34	0.31	10	96.3
9	Hemisphere	8.14	0.17	15	97.9

TABLE 6.7
Dimensional Deviation of FFF Parts, Measured before and after Treatment

Test Number	Form	Average Dimensions: Before Treatment (mm)	Average Dimensions: After Treatment (mm)	Treatment Time (s)	Dimensional Deviation (%)
1	Cube	*h*: 25.80	*h*: 25.75	10	0.19
2	Cube	*h*: 25.60	*h*: 25.53	15	0.27
3	Cube	*h*: 25.89	*h*: 25.45	20	1.69
4	Cylinder	*h*: 23.69 ϕ: 29.96	*h*: 23.63 ϕ: 29.82	15	0.46
5	Cylinder	*h*: 23.76 ϕ: 29.93	*h*: 23.71 ϕ: 29.82	20	0.36
6	Cylinder	*h*: 23.82 ϕ: 29.92	*h*: 23.76 ϕ: 29.83	10	0.30
7	Hemi-sphere	*h*: 20.41 ϕ: 39.86	*h*: 20.30 ϕ: 39.77	20	0.22
8	Hemi-sphere	*h*: 20.42 ϕ: 39.80	*h*: 20.31 ϕ: 39.76	10	0.10
9	Hemi-sphere	*h*: 20.41 ϕ: 39.82	*h*: 20.25 ϕ: 39.77	15	0.12

h is the height of the part and ϕ is the diameter of the part [26].

and understood by the industry. As it has already been applied to products that we use in our everyday life. Some examples of industries that already use electroplating for surface finish enhancement are automotive, jewelry, aircraft industry, and much more. Electroplating methods can provide properties that can't be obtained with other methods. Such as for automotive applications, polymer parts are electroplated for

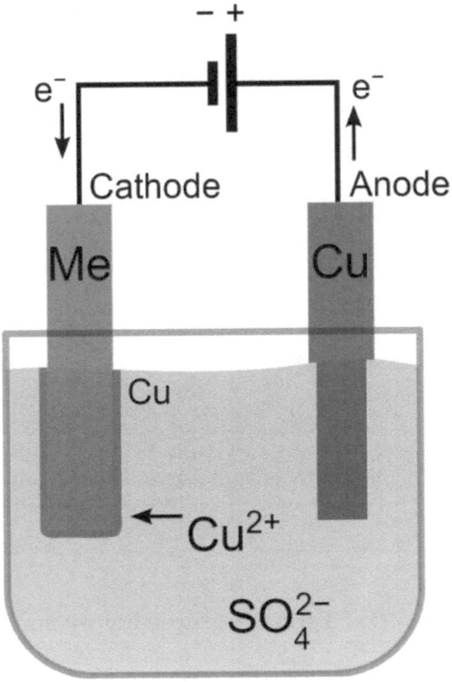

FIGURE 6.6 Basic representation of the electroplating process [32].

better mechanical properties and superior surface finish. Further, the following properties can be tuned using electroplating post-processing:

- increasing Young's modulus;
- preventing corrosion (for metals);
- increasing the conductivity;
- increase chemical resistance.

Since most polymers are non-conductive, and thus not suitable for being used as a cathode in the electroplating process. A different approach called electroless plating is needed to deposit a metallic coating on polymer parts. The electroless plating process involves a polymer substrate that is submerged in a bath similar to the electroplating process. However, in contrast to electroplating, there is no need to apply electrical currents or use electrodes. The deposition of solid-phase coatings is achieved through a series of complex chemical reactions instead of just one electrochemical reaction [30]. The overall basis of the electroless deposition process is represented by the following reaction [31]:

$$\text{Reductant} + M^{Z+} \underset{\text{Catalytic surface}}{\longrightarrow} \text{Oxidation product} + M \downarrow \tag{1}$$

The bath in electroless plating is very complex and usually contains multiple components, such as [30, 31]:

- metal salt
- chelator
- stabilizer
- a reduction agent
- pH adjustment agent
- accelerator
- buffer
- wetting agent

The metal salt will function as a provider for the metal ions (M^{Z+}), these metal ions will react with the reducing agent. The reducing agent or reductant donates its electrons and thus reducing the metal ions into their metallic form. Additionally, the reducing agent can also be used to provide non-metallic elements that might be required if a specific alloy plating is desired. If only the metal ions with the reductant are present in the solution, precipitation of the metal (M) and metal salts will occur. To prevent the immediate precipitation of metal and metal salts in the bulk solution, the use of a chelator is advised. A chelator's role is to complex the metal ions and prevent the occurrence of a free metal-ion concentration [30, 31]. On top of this, complexing agents can work as pH buffers as well. Another component that helps keep the solution stable, is the stabilizer. The stabilizer will inhibit catalytically active particles, thus stopping the spontaneous decomposition of the electroless solution. However, an excess of stabilizer will result in a decrease or even stop the plating rate. For this reason, it is best to only use a trace amount of this stabilizer. The counterpart of the stabilizer is the accelerator, which will activate the reducing agent and thus accelerate the deposition of the metal film. To assure a fine and even deposition of the metal film, the use of a wetting agent is advised. The wetting agent increases the wettability of the substrate that is to be coated. This means that the contact angle between the surface and the solution will decrease and thus aiding the escape of H_2 gas. This is needed because, the formation of H_2 gas, or any other gaseous substance for that matter, will increase the porosity of the deposit [30, 31]. As shown in Figure 6.7 once the part has gone through the electroless deposition process, it can be subjected to electroplating to apply additional metal layers.

According to a recent research study [34], Young's modulus of an SLA additive manufactured part increases with the thickness of the electroplated coating, as can be seen in Figure 6.8. A higher degree of resistance to external stress is a natural result of the metallic coating on the surface of the part and is directly proportional to the thickness of the coating.

Similar results were also found for the ultimate tensile strength, elongation, and impact strength as shown in Figure 6.9.

6.2.3 Polymer Coatings

The application of polymer coatings as a chemical post-processing method for AM processes is a great way to improve the outer – as well as the inner surface quality of AM parts. Since this is a non-destructive method, good surface properties can be achieved without a loss in dimensional accuracy [35]. Polymers are macromolecules

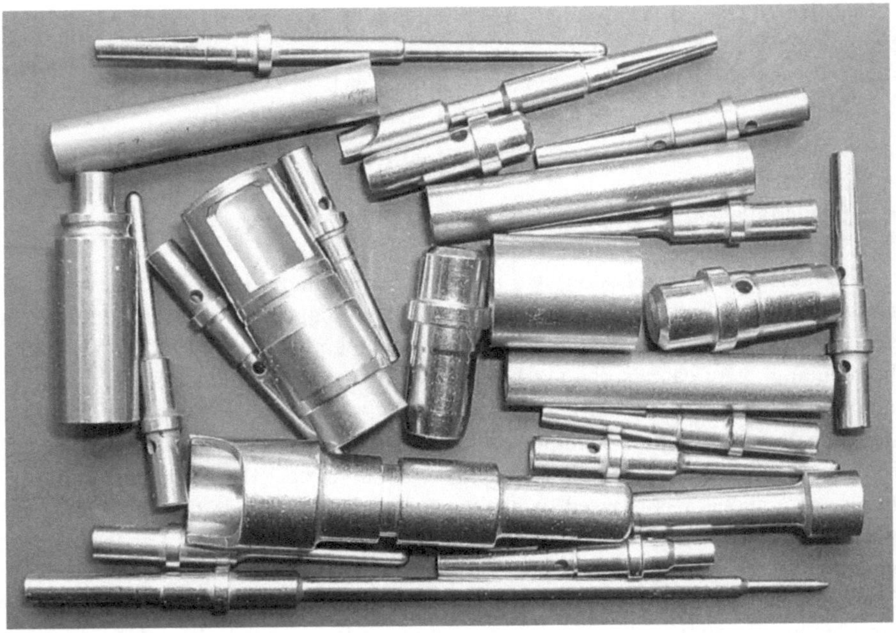

FIGURE 6.7 Electroplated products [33].

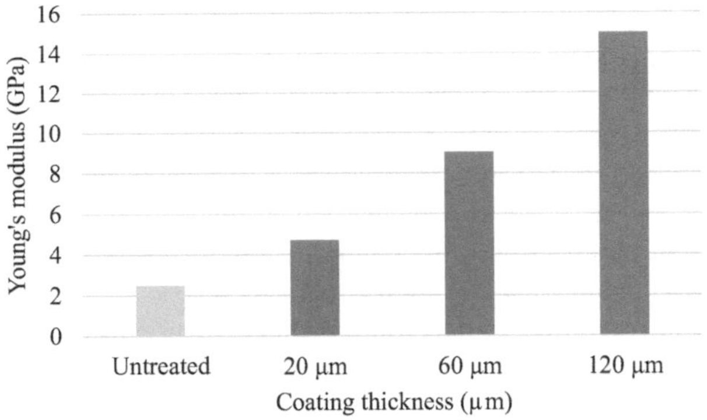

FIGURE 6.8 Comparison of Young's modulus between different coating thicknesses.

that consist of many low molecular weight units called monomers, these are linked together by covalent bonds and a simple representation of the process can be seen in Figure 6.10.

Due to the flexible nature of coatings, it is possible to improve or even add different properties to AM parts. This is mainly achieved by adding different additives. Some of the properties that can be improved or added are as follows:

- UV-resistance;
- anti-microbial;

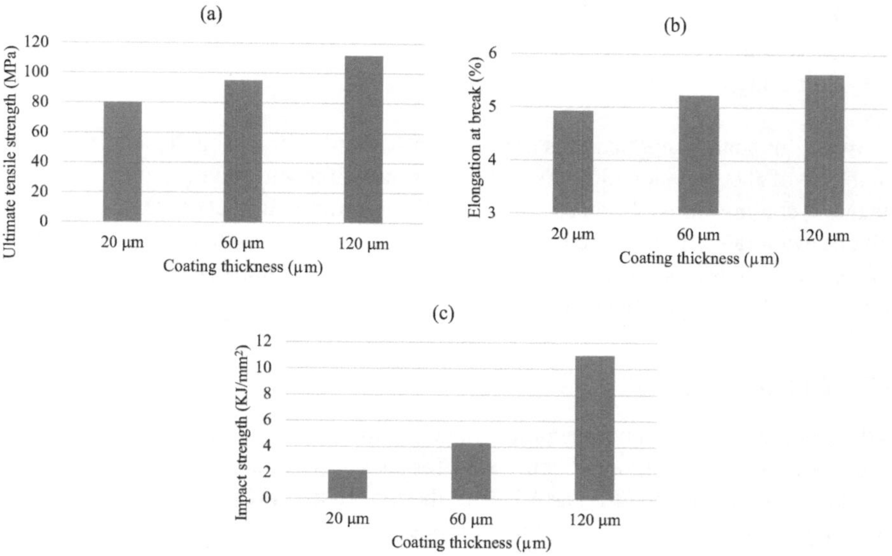

FIGURE 6.9 (a) Ultimate tensile strength comparison between different layer thicknesses, (b) elongation at break comparison between different layer thicknesses, and (c) impact strength comparison between different layer thicknesses.

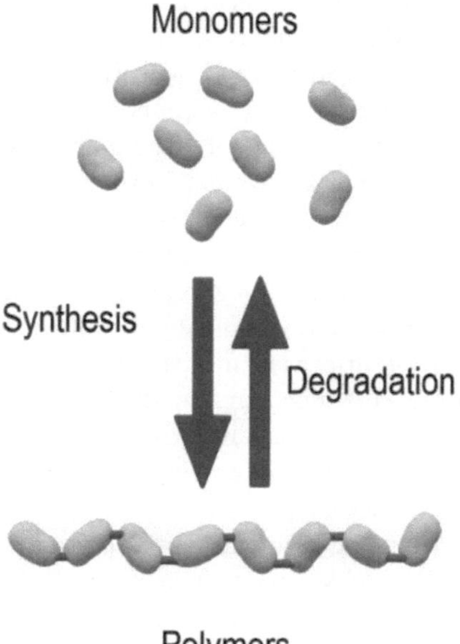

FIGURE 6.10 Basic reaction scheme for polymers forming out of monomers [38].

- hydro- or lipophilicity;
- antistatic;
- flame suppressing.

Most film coating applications are solution or dispersion-based, depending on the solubility of the polymer [36]. As of today, spray atomization is the most common means of coating films. When sprayed onto the substrate, droplets spread, and as the solvent evaporates, polymer chains interpenetrate and form the coating [37]. Once drying is complete, a smooth surface is formed and thus eliminates the staircase effect and shields the AM part beneath it.

6.3 CONCLUSIONS

In the last few years, AM has become a standard practice in the design and fabrication of many industrial components and even for mass-production applications. Out of the many AM processes FFF and VPP are the most used AM processes due to many advantages, such as different polymer/composite material options, low cost, and easy availability. One of the major challenges of the FFF and VPP manufactured parts is the presence of surface roughness due to the process characteristics, i.e., layer-by-layer material addition (stair-casing effect). And as these processes have made their way to the customer-grade products, it is important to investigate and enhance their surface properties along with their mechanical properties. Among other post-processing methods to enhance surface finish, chemical post-processing is a less traveled path, especially in the AM community.

This chapter introduces the reader to FFF and VPP processes, different material options, and various chemical post-processing options that can be utilized in these processes. Starting with a basic introduction to the processes, the chapter dives right into the chemical post-processing methods and explains the need for it, its various types, and possible advantages, the surface finish of the printed part being the most important of those. Moving on, the chemical vapor smoothing of the polymer AM parts is introduced and explained in detail along with the quantitative surface roughness improvement. Further, electro/electroless plating is explained especially for polymer AM parts and its effect on the surface and mechanical properties is discussed. Finally, a polymeric coatings-based post-processing method is introduced for polymer AM parts and the potential use cases and advantages are explained. It is one of the only post-processing methods that can enhance many other special properties of the part along with the surface properties.

Overall, this chapter covers the different chemical-based post-processing methods that can be utilized for polymer AM parts to increase the surface and other properties. The readers are welcome to explore the references for further readings and detailed knowledge.

REFERENCES

[1] Gibson I, Rosen DW, Stucker B, Khorasani M, Rosen D, Stucker B, Khorasani M. Additive manufacturing technologies. Cham, Switzerland: Springer; 2021.

[2] Kumar S, Singh R, Singh TP, Batish A. Fused filament fabrication: A comprehensive review. Journal of Thermoplastic Composite Materials. 2023. 36: 794–814.
[3] Pagac M, Hajnys J, Ma QP, Jancar L, Jansa J, Stefek P, Mesicek J. A review of vat photopolymerization technology: Materials, applications, challenges, and future trends of 3D printing. Polymers. 2021. 13: 598.
[4] Wikimedia commons [Online], Available: https://commons.wikimedia.org/wiki/File:3D_Printer_Extruder.png Accessed on: 08 Jan. 2023
[5] Elkaseer A, Schneider S, Scholz SG. Experiment-based process modeling and optimization for high-quality and resource-efficient FFF 3D printing. Applied Sciences. 2020. 10: 2899.
[6] Dey A, Roan Eagle IN, Yodo N. A review on filament materials for fused filament fabrication. Journal of Manufacturing and Materials Processing. 2021. 5: 69.
[7] Rupal BS, Ramadass K, Qureshi AJ. Investigating the effect of motor micro-stepping on the geometric tolerances of Fused Filament Fabrication printed parts. Procedia CIRP. 2020. 92: 9–14.
[8] Harris M, Potgieter J, Archer R, Arif KM. Effect of material and process specific factors on the strength of printed parts in fused filament fabrication: A review of recent developments. Materials. 2019. 22: 12.
[9] Kumar S, Singh R, Singh TP, Batish A. Fused filament fabrication: a comprehensive review. Journal of Thermoplastic Composite Materials. 2020. 36: 794–814.
[10] Zanjanijam AR, Major I, Lyons JG, Lafont U, Devine DM. Fused filament fabrication of PEEK: A review of process-structure-property relationships. Polymers. 2020. 12: 1665.
[11] Mohamed OA, Masood SH, Bhowmik JL. Optimization of fused deposition modeling process parameters: A review of current research and future prospects. Advances in Manufacturing. 2015. 3: 42–53.
[12] Zhang F, Zhu L, Li Z, Wang S, Shi J, Tang W, Li N, Yang J. The recent development of vat photopolymerization: A review. Additive Manufacturing. 2021. 48: 102423.
[13] Al Rashid A, Ahmed W, Khalid MY, Koç M. Vat photopolymerization of polymers and polymer composites: Processes and applications. Additive Manufacturing. 2021. 47: 102279.
[14] Pazhamannil RV, Govindan P. Current state and future scope of additive manufacturing technologies via vat photopolymerization. Materials Today: Proceedings. 2021. 43: 130–136.
[15] Huang J, Qin Q, Wang J. A review of stereolithography: Processes and systems. Processes. 2020. 8: 1138.
[16] Rupal BS, Ahmad R, Qureshi AJ. Feature-based methodology for design of geometric benchmark test artifacts for additive manufacturing processes. Procedia CIRP. 2018.
[17] Singh S, Singh G, Prakash C, Ramakrishna S. Current status and future directions of fused filament fabrication. Journal of Manufacturing Processes. 2020. 70: 84–89.
[18] Kumbhar NN, Mulay AV. Post processing methods used to improve surface finish of products which are manufactured by additive manufacturing technologies: a review. Journal of The Institution of Engineers (India): Series C. 2018. 99: 481–487.
[19] Zhu Z, Keimasi S, Anwer N, Mathieu L, Qiao L. Review of shape deviation modeling for additive manufacturing. In Advances on Mechanics, Design Engineering and Manufacturing 2017 (pp. 241–250). Springer, Cham.
[20] Lavecchia F, Guerra MG, Galantucci LM. Chemical vapor treatment to improve surface finish of 3D printed polylactic acid (PLA) parts realized by fused filament fabrication. Progress in Additive Manufacturing. 2022. 7: 65–75.

[21] Miller-Chou BA, Koenig JL. A review of polymer dissolution. Progress in Polymer Science. 2003. 28: 1223–1270.

[22] Singh R, Singh S, Singh IP, Fabbrocino F, Fraternali F. Investigation for surface finish improvement of FDM parts by vapor smoothing process. Composites Part B: Engineering. 2017. 111: 228–234.

[23] Xu K, Xi T, Liu C. Design of the desktop vapor polisher with acetone vapor absorption mechanism. In Journal of Physics: Conference Series 2019 Aug 1 (Vol. 1303, No. 1, p. 012061). IOP Publishing.

[24] Kuo CC, Mao RC. Development of a precision surface polishing system for parts fabricated by fused deposition modeling. Materials and Manufacturing Processes. 2016. 31: 1113–1118.

[25] Kanani N. Electroplating: basic principles, processes and practice. Elsevier; 2004.

[26] Zhang BW, Liao SZ, Xie HW, Zhang H. Progress of electroless amorphous and nano alloy deposition: a review–Part 2. Transactions of the Institute of Metal finishing. 2014. 92: 74–80.

[27] Djokić SS, Cavallotti PL. Electroless deposition: theory and applications. Electrodeposition. 2010. 48: 251–289.

[28] Wikimedia commons [Online], Available: https://commons.wikimedia.org/wiki/File:Copper_electroplating_principle_(multilingual).svg Accessed on: 08 Jan. 2023

[29] Wikimedia commons [Online], Available: https://commons.wikimedia.org/wiki/File:Gold-plated_electrical_connectors.jpg Accessed on: 08 Jan. 2023

[30] Saleh N, Hopkinson N, Hague RJ, Wise S. Effects of electroplating on the mechanical properties of stereolithography and laser sintered parts. Rapid Prototyping Journal. 2004. 10: 305–315.

[31] Yang Q, Lu Z, Zhou J, Miao K, Li D. A novel method for improving surface finish of stereolithography apparatus. The International Journal of Advanced Manufacturing Technology. 2017. 93: 1537–1544.

[32] Cuppok Y, Muschert S, Marucci M, Hjaertstam J, Siepmann F, Axelsson A, Siepmann J. Drug release mechanisms from Kollicoat SR: Eudragit NE coated pellets. International Journal of Pharmaceutics. 2011. 409: 30–37.

[33] Kurakula M, Naveen NR, Yadav KS. Formulations for Polymer Coatings. Polymer Coatings: Technology and Applications. 2020.

[34] Wikimedia commons [Online], Available: https://upload.wikimedia.org/wikipedia/commons/6/6f/Synthesis_and_Degradation.png Accessed on: 08 Jan. 2023

7 Coating/Cladding Based Post-Processing in Additive Manufacturing

Rashi Tyagi[1,2] *and Ashutosh Tripathi*[2]
[1]Department of Mechanical Engineering, Chandigarh University, Gharuan, Punjab, India
[2]University Center for Research and Development, Chandigarh University, Gharuan, Punjab, India

CONTENTS

7.1 INTRODUCTION

Additive manufacturing (AM) has several benefits, such as preparation of intricate geometrical shapes, ability to form lattice structures inside a component to reduce its weight, less wastage of material, and less cost. Nevertheless, an additive process has a limitation that most of the time the prepared part exhibits high surface roughness, which makes the AM process difficult to use in mass production [1]. This high surface roughness will be a negative factor in various industrial applications. Aerodynamic performance, e.g., for turbine blades or vanes through interaction of the boundary layer of air flowing along the surface and resulting into flow instabilities [2]. Therefore, a post-processing of the additive part is generally performed to reduce the roughness. In addition, post-processing of the AM part facilitates improvement in other material properties.

DOI: 10.1201/9781003276111-7

There are two approaches for post-processing the additive part: (1) removing the material from a part, (2) post-treating a part using the additive approach by coating deposition over the part. Additive based post-treatments have various advantages over subtractive-based post-treatments. Here, the additive-based approach does not need the removal of material, leading to unchanged mechanical properties. Additionally, no direct contact takes place between the machine used for coating and the part, while use of subtractive processes for post-treatment is hard for an intricately shaped part and also consumes time. However, post-treatments using the additive approach utilizing commercial coating methods requires a high amount of material deposition onto the AM part such that the part dimensions are considerably increased. Additionally, the study the adhesion properties between coating and the AM part surface is an important aspect while using additive-based post-treatment techniques [3]. Surface coating is one of the post-processing processes in the three-dimensionally (3D) printed part. In addition, surface coating facilitates improvement in any desirable material properties. This study explains the capability of surface coating processes to improve the quality and material properties of the additively manufactured part. In this work, much attention has been paid to spraying processes, sputtering, and electroless plating process.

7.2 RESEARCH GAP

Table 7.1 shows a bibliographic study, which is performed to generate a bibliographic map made with the help of a VOS viewer. By setting the minimum number of occurrences as 9, a total of 99 terms have fulfilled the minimum criteria according to the VOS viewer out of the total terms of 8398. Then, the VOS viewer suggested 59

TABLE 7.1
Relevance Score and Occurrences of 33 Terms Best Suitable for the Present Work

ID	Term	Occurrences	Relevance Score
1	3D printer	22	0.7022
2	ABS	20	0.5238
3	additive manufacturing	38	0.811
4	adhesion	24	0.6199
5	antenna	37	1.1414
6	coating	49	0.5035
7	copper	56	0.4812
8	development	32	0.4122
9	effect	22	0.5481
10	electroless plating	60	0.2262
11	electroplating	17	0.4859
12	FDM	14	0.5369
13	feasibility	21	1.0983
14	formation	18	0.5038
15	implementation	12	2.0179

(*Continued*)

TABLE 7.1 (Continued)
Relevance Score and Occurrences of 33 Terms Best Suitable for the Present Work

ID	Term	Occurrences	Relevance Score
16	influence	11	1.3371
17	investigation	19	0.5712
18	limitation	15	3.4008
19	measurement	33	0.5403
20	mechanical property	16	1.5103
21	metallization	37	0.3624
22	nickel	22	0.8111
23	PLA	14	0.8926
24	polymer	43	0.2482
25	prototype	15	0.552
26	sensor	20	0.9761
27	shape	19	1.2732
28	silver	17	1.0276
29	simulation	23	0.6306
30	SLA	15	4.4308
31	stereolithography	27	2.6403
32	strength	34	0.8718
33	surface roughness	12	0.3114

terms that are best suitable. Finally, 33 terms are manually selected as per the present research work. Figure 7.1 shows these 33 terms in map having various clusters (green, red, yellow, and blue). Different terms related to different areas are categorized within different cluster. Red color shows cluster 1, green color shows cluster 2, blue color shows cluster 3 and has 12, 12, and 5 terms, respectively. As per Table 7.1, Figures 7.1 and 7.2 are obtained by the VOS viewer that illustrates past research work done related to coating, plating, and metallization processes.

Figure 7.2 explains the research gap between additive manufacturing and all the terms related to the present work done to date. As per Figure 7.2, there are lots of research work that is performed on coating and electroless plating. Figure 7.2 also shows research work that is less reported, such as cold spray coating, dip coating, and sputtering processes.

7.3 DIFFERENT PROCESSES OF COATING

7.3.1 Spray Coating

Spray coating is mainly applied for painting car body or many other applications for which thick coatings of 100 μm are deposited on a freeform product. Figure 7.3 shows the deposition process and curing before formation of texture with fluorinated silica nanoparticles. In this process, a compressed air in the form of spray and gravity feed cup is utilized, which aerosolize the suspensions in order to prepare coating. Then, a

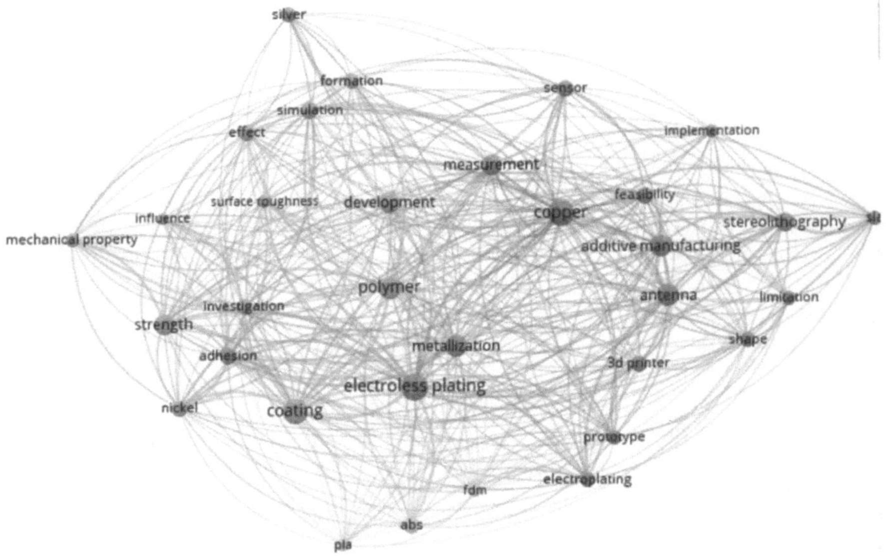

FIGURE 7.1 Bibliographic study of coatings and its applications.

FIGURE 7.2 Bibliographic study of additive manufacturing with different terms.

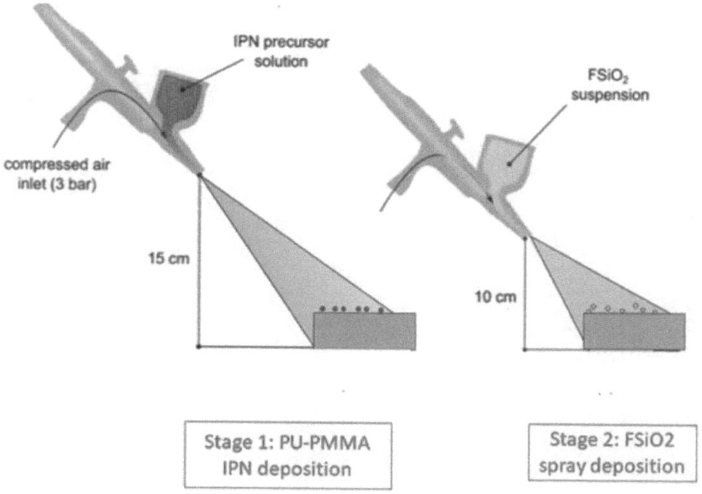

FIGURE 7.3 Arrangement of the two-stage coating prepared by spray deposition [4].

particular value of compressed air pressure is set to drag the feed liquid at a required rate of flow. Kreider et al. investigated the performance of superhydrophobic coatings of fluorinated silica to prevent the degradation of 3D-printed polyamide composites/carbon-fiber, which was caused by the moisture. Owing to the superhydrophobic coating formation over the 3D-printed polyamide, the presence of liquid water repelling behavior in the surface features diminished the moisture-induced swelling of the PA matrix. Additionally, coated PA showed tensile yield strength ranging from 6.7–12.4%, greater than the without coating PA. Also, by optimizing the coatings it is possible to reach a solution to improve the properties of PA composite in both wet and humid environments [4].

The conventional-type additive post-processing, e.g., as pneumatic spray coating, needs a high amount of material to be coated. In contrast, ultrasonic spray coating technology for additive post-treatment has the ability to apply coating in a productive way, leading to less material usage. Ultrasonic spray coating (USSC) process is carried out via ultrasonic atomization of ink. The working principle of USSC is shown in Figure 7.4 [3]. Steirer et al. (2009) showed that, as compared to p-xylene, the films sprayed from chlorobenzene led to higher efficiencies of device because the film morphologies are very different [5].

To avoid the risk of damage, cold additive spray technology (CSAT) is an emerging spray technology in recent times due to the possibility to obtain thick coatings. CSAT utilized high-pressure gas, e.g., N_2, air, or He, under supersonic conditions to push and transfer metal powder with particle size 10–100 μm over the substrate. CSAT can deposit high-strength coating and can form 3D parts on the workpiece. This supersonic effect of metal particles results in primarily mechanical interlocking and formation of partial metallurgical bond under solid-state surroundings with negligible melting the powders. Compared to other coating processes, cold spray is the most desirable technique for metallization of materials

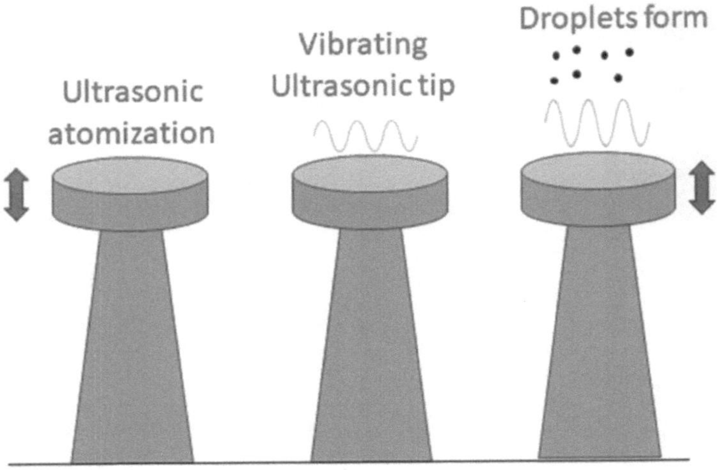

FIGURE 7.4 Schematic of ultrasonic atomization process of the ink [3].

that are temperature-sensitive, e.g., composites and polymers with thermoplastic matrix. This method is based on the kinetic energy for the coating formation instead of thermal energy and subsequently degradation and erosion of polymer-based composite could be avoided. Denzy et al. (2021) used onyx as a workpiece material, which is composed of nylon, a thermoplastic matrix, and chopped carbon fibers randomly dispersed in it. In recent times, fused filament fabrication (FFF) is emerging in industrial applications, which is a 3D-printing process that requires thermal extrusion of thermoplastic polymers in the form of filaments from a heated nozzle. This technique leads to production of customized composites for the cold spray applications. The cold spraying of aluminum powder was done on the Onyx under a low-pressure cold spray system [6]. Table 7.2 summarizes the various outcomes of the spraying process in 3D printing.

7.3.2 Dip Coating

The process of dip coating results in the best surface finish, particularly for complex parts, which will be suitable to obtain the improved surface of additive prepared coated sand mold [1]. This process is mostly employed to enhance the surface quality of coated sand mold prepared by additive manufacturing [1]. Yao et al. (2016) used a dip coating process to eliminate the staircases and high surface roughness issue. This issue can be eliminated by coating of alcohol and water-based refractory coating. Alcohol-based coating leads to increasing coating thickness speedily and decreasing tensile strength with the dipping time. These results indicated that dipping time will be limited due to requirement of coating thickness and strength. In water-based coating, a constant coating thickness and tensile strength of molds with dipping time is maintained. This showed that there would be sufficient time to dip to assure the coating uniformity. Also, thickness of coating can be controlled by modifying the wet

TABLE 7.2
Outcomes of Spraying Process in 3D Printing

Additive Manufacturing Process	Base Material (Material of 3D-Printed Part)	Coating Material	Outcome	Reference
Fused filament fabrication (FFF)	carbon fiber-reinforced plastics (CFRP) panels	Aluminum particles	• Surface coverage close to 100% • Adhesion strength (close to 4 MPa) • Less distortion during the deposition	[7]
FFF	ONYX–Based composites	Aluminum Powder	• No void formation • Increase of the coverage	[8]
Steriolithography (SLA)	Isomalt and Hydrogels (water soluble polymer)/	octadecane	• To generate complex channels within hydrogels • To obtain biocompatible properties	[9]
SLA	polyamide 12	polyvinylidene fluoride (PVDF)	• Roughness reduction was achieved	[10]
FFF	PLA	Aluminum spray	• Improvement in compressive strength	[11]
FFF	Onyx made of nylon mixed with short carbon fibers.	Onyx polymeric matrix and aluminum powders	• To prepare 3D-printed PMCs panels	[12]
FFF	s Onyx, a novel nylon-based material with short carbon fibers dispersed in it	Aluminum powder	• Good surface coverage	[13]

coating density, yield stress, and wet dry coating [1]. A water-based antimicrobial coating for 3D-printed parts was discussed by Zhu et al. (2015) and they showed less bacterial growth than without treated printed parts. They explained that it was possible to print medical devices without changing existing 3D printers [14]. Camović et al. (2020) used dip coating method for 3D printing of microneedles, which can be used in drug delivery. Biodegradable 3D-printed PLA microneedles are an emerging class of novel transdermal drug delivery systems. They showed that 3D printing along with the post-fabrication etching can make it possible to obtain ideal size and shaped microneedles [15]. Table 7.3 shows the outcomes of electroless plating in 3D printings.

TABLE 7.3
Outcomes of Dip Coating Process in 3D Printing

Additive Manufacturing Process	Base material (Material of 3D-Printed Part)	Coating Material	Outcome	Reference
FDM	ABS structure	Rubber was diluted with toluene	• To improve the adhesion of superhydrophobic coating	[16]
FDM	PLA	Silica nano particles	• To obtain superhydrophobic surface • For low water adhesion applications	[17]
FDM	ABS, PLA	borne alkyd coating, water borne acrylic coating and coating of ABS	• Mechanical properties were obtained at differently compositions of coatings • Surface roughness was reduced	[18]
FDM	carbon-based electrode	MXene ($Ti_3C_2T_x$) and solid lubricants, eg, MoS_2, $MoSe_2$, WS_2, and WSe_2	• Improvement in catalytic properties of nanocarbon electrodes • For energy conversion applications.	[19]
FDM	PLA scaffold	Polydopamine (PDA) containing gold salt	• Improved cell adhesion	[20]
FDM	PLA	silica nanoparticles	• Hydrophobic properties were achieved with contact angle>150°	[21]

7.3.3 Electroless/Electro Plating

The electroplating process is a simple process that is recently used in many sectors, including electronic manufacturing and jewelry and for the preparing the metal layers and finishes. The plating process involves a current flowing between two electrodes, which is termed as a cathode and an anode. During this process, these electrodes are submerged in an electrolyte bath. Electroplating can be done to enhance the strength of a part by improving Young's modulus and durability, corrosion. Also, the conductivity of the part can also be improved by plating.

There are many electronic devices prepared by 3D printing. While its applications are limited by the electrically insulating characteristics of polymers. Here, the part to be coated must be conductive in electroplating, because the metal ions are deposited by traveling from anode to cathode in order to form a coating layer. 3D-printing technology utilizes polymers, their post-processing needs to be done before performing the electroplating process on them. Therefore, the polymer parts need a conductive coating, such as by applying silver paint/copper and nickel filled or any other conductive paints, roughening, and by electroless plating of conductive materials. Other simple post-processing of plastic parts to make them conductive involves rubbing action by using graphite or spraying of silver [22]. Coating on these parts, i.e., antenna in metal, leads to improvement in conductivity for lower resistance. Therefore, electroplating over the non-conductive parts by carrying out these post-processing approaches have been used on 3D-printed parts fabricated by using, laser sintering, stereolithography, and FFF-printed parts.

The methods of making the conductive surface have limitations because the entire surface is to be coated. In this regard, a different way of plating a non-conductive part is adopted by Angel et al. (2018) in which they elaborated that electroplating of a part can be done directly from a 3D printer by using both conductive and non-conductive filament. They showed FFF-printed parts can be plated to enhance the electrical properties and to print electromechanical systems. In this regard, both non-conductive and conductive filament can be electroplated by 3D printer using the conductive filament to point out the location where the metal deposition will take place. The other various isolated conductive segments were added to plate separately to permit the addition or accumulation of multiple plating materials on the same part [23]. However, the plating process generally needs an activation step with expensive catalysts prior to electroless plating, and the bath used in electroless plating is chemically not stable, which needs to be replaced regularly. Table 7.4 shows the outcomes of electroless plating in 3D printing.

7.3.4 Magnetron Sputtering

In the sputtering process, bombardment of high-energy particles causes the atoms from the material to come out from a target. Figure 7.5 shows the process of sputtering. A–D shows the four sizes of tetrahedral truss structures that is printed over the glass as given in Figure 7.5a. Figure 7.5b and 7.5c shows the configurations of two sputtering that is used. In one configuration, samples were sputtered using a stationary setup, where the sample holder is in direct line-of-sight of the target and actively cooled from the backside using flowing water at 16 °C (Figure 7.5b). The second setup uses a holder that rotates at 54 rpm throughout sputtering (Figure 7.5c) without cooling [30]. Magnetron sputtering is highly efficient vacuum coating process to deposit the metals, alloys, and compounds over the variety of materials having mm thicknesses. It has various benefits as compared to other vacuum coating process, such as production of a high number of commercial applications from microelectronic to simple decorative coating. Also, many works have been reported on the better adhesion of metallic coating over the various types of workpieces via sputtering processes. Afshar et al. (2020) reported that the metallization process to coat structures is an interesting work to improve the electrical and structural properties of printed polymers. Metallization can be done by different processes, such as dipping, electroless deposition, thermal evaporation, electric arc spraying, and stereolithography. These processes have limitations that they cannot precisely control the metals' deposition and also needs high temperature surroundings, use of harsh chemicals and preparation of surface,

TABLE 7.4
Outcomes of Electroless Plating in 3D Printing

Additive Manufacturing Process	Base Material (Material of 3D-Printed Part)	Coating Material	Outcome	Reference
Direct laser writing	artificial bacterial flagella	CoNiP	• Biomedical applications	[24]
FDM and SLA	ABS composite	palladium	• Chemical and mechanical engineering systems	[25]
SLA	Photocurable resin	Cu and NiP	• Cu and NiP were successfully deposited on the resins • Can be used as mechanical actuators for the functioning of other structures, for electrical switching or microfluidic control.	[26]
FDM	Polyactic and Cu reinforced Composite structure	Copper	• Faradaic peak separation value (70–75 mV) is obtained after coating • To prepare mesh electrodes for spectro electrochemical applications	[27]
FDM	ABS	Copper	• Preparation of EDM electrode by Cu plating the ABS	[28]
FDM	ABS	Copper	• Mentalization of ABS Part • Better electrical performance.	[29]

which could be non-compatible with the workpiece or negate the recyclability of thermoplastic materials. However, the process of magnetron sputtering allows a precision coating process under low temperature while not needing pre-treatment of structural surface and without any harmful impact on environmental with reduced volume and weight [31].

Many researchers used magnetron sputtering process to enhance the tribological properties of additive manufactured part. Tillman et al. (2020) deposited DLC coatings by using sputtering process to significantly enhance the tribological properties of 36NiCrMo16. During the tribo testing, the DLC coatings showed enhancement in tribo-mechanical properties of SLM 36NiCrMo16 substrates, and an excellent concept to attain simultaneous reduction in weight in highly stressed parts for small batch production [32]. Juarez et al. (2018) used magnetron sputtering process to fabricate coating of three different metals on the microcellular structures prepared with 3D direct laser writing [30]. They performed compression tests that revealed that mechanical properties enhanced with coating material [30]. Sputtering et al.

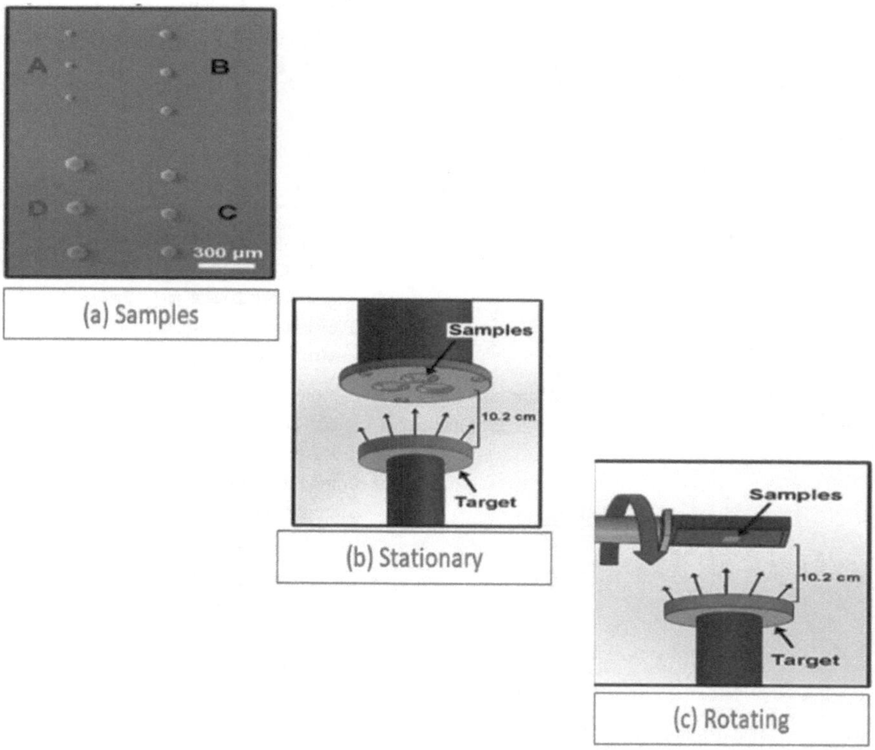

FIGURE 7.5 Experimental setup for sputtering shows (a) SEM image of layout of samples on the glass slide. These samples are labeled as per the increasing strut length [*A*=5 µm, *B*=7.5 µm, *C*=10 µm, and *D*=12.5 µm], (b) depicts the representation of sputtering for stationary sputtering, (c) shows the rotating sputtering setup with sample mounted at the center of the rotating holder. Sample is rotated once every 54 seconds [30].

(2019) applied radio-frequency sputter coating on 3D cranial meshes of Ti6Al4V prepared from selective laser melting (SLM). Titanium-based implants, has one of the drawbacks of slow osseointegration in spite of their attractive biocompatibility and mechanical properties. RF-MS coating was applied for biological performance [33]. However, addition of coating can significantly enhance the osseointegration if the material used in the manufacturing of metallic implants is bioinert, obtaining uniform/homogenous coatings of the surfaces for the parts having complex 3D structure, and mainly the ones exhibiting porous 3D-lattice structures is very difficult. In this direction, Chudinova et al. (2016) used RF-magnetron sputtering on Ti64 scaffolds to deposit biocompatible hydroxyapatite (HA) coating [34]. They showed that additively manufactured metallic 3D-lattice structures can be effectively coated homogeneously with hydroxyapatite using magnetron-based deposition process [34]. Table 7.5 shows the outcomes of sputtering in 3D printing.

TABLE 7.5
Outcomes of Sputtering in 3D Printing

Additive Manufacturing Process	Base Material (Material of 3D-Printed Part)	Coating Material	Outcome	Reference
FDM	ABS	Copper	• Enhanced durability of structure • Improved mechanical properties • Prevention against harsh environmental conditions	[31]
Direct laser writing	epoxy	Aluminum Ti-6Al-4V, and Inconel	• Development of developing metal-polymer tetrahedral trusses • Improvement in conductivity, strength, and chemical activity was achieved	[30]
FDM	ABS	**CrN**	• Defect-free surface is obtained which is bright and reflective and electrochemically and chemically stable	
FDM	ABS and PLA	gold thin film	• Cell adhesion, proliferation • Better wettability	[35]
FDM	ABS	Copper	• Good and corrosion resistance by maintaining the flexural strength, ductility and toughness	[36]

7.4 CONCLUSIONS

This study reported the coating of an additively manufactured part to spread its scope into industrial applications. The following are the conclusions obtained from this study:

- The process of cold spray is found to be most suitable for the metallization of materials that are temperature-sensitive, e.g., composites and polymers with a thermoplastic matrix.
- Conventional techniques of post-treatments of additive part, for example, pneumatic spray coating, needs a high volume of coating material. However, ultrasonic spray coating process has the ability to prepare coating in the most efficient way, leading to less material usage.
- The electroless plating method needs step of activation and costly catalysts, and the bath used in electroless plating is chemically not stable, which needs to be replaced regularly.

- Dip coating process is mostly used to improve the surface quality of complex shaped parts. The part prepared by additive manufacturing exhibit staircases on the surface. This staircase effect could be enhanced by both water-based and alcohol-based refractory dip coating.

REFERENCES

[1] S. Yao, T. Wang, Improved surface of additive manufactured products by coating, J. Manuf. Process. 24 (2016) 212–216.

[2] P.D. Enrique, A. Keshavarzkermani, R. Esmaeilizadeh, S. Peterkin, H. Jahed, E. Toyserkani, N.Y. Zhou, Enhancing fatigue life of additive manufactured parts with electrospark deposition post-processing, Addit. Manuf. 36 (2020) 1–22. https://doi.org/10.1016/j.addma.2020.101526.

[3] S. Slegers, M. Linzas, J. Drijkoningen, J. D'Haen, N.K. Reddy, W. Deferme, Surface roughness reduction of additive manufactured products by applying a functional coating using ultrasonic spray coating, Coatings. 7 (2017) 208.

[4] P.B. Kreider, A. Cardew-1, S. Sommacal, A. Chadwick, S. Hümbert, S. Nowotny, D. Nisbet, A. Tricoli, P. Compston, The effect of a superhydrophobic coating on moisture absorption and tensile strength of 3D-printed carbon-fibre/polyamide, Compos. Part A Appl. Sci. Manuf. 145 (2021) 106380.

[5] K.X. Steirer, M.O. Reese, B.L. Rupert, N. Kopidakis, D.C. Olson, R.T. Collins, D.S. Ginley, Ultrasonic spray deposition for production of organic solar cells, Sol. Energy Mater. Sol. Cells. 93 (2009) 447–453. https://doi.org/10.1016/j.solmat.2008.10.026.

[6] R. Della Gatta, A. Astarita, D. Borrelli, A. Caraviello, F. Delloro, P. Lomonaco, I. Papa, A.S. Perna, R. Sansone, A. Viscusi, manufacturing of aluminum coating on 3D-printed onyx with cold spray technology, (2021).

[7] A.S. Perna, A. Viscusi, R. Della Gatta, A. Astarita, Integrating 3D printing of polymer matrix composites and metal additive layer manufacturing: surface metallization of 3D printed composite panels through cold spray deposition of aluminium particles, Int. J. Mater. Form. 15 (2022). https://doi.org/10.1007/s12289-022-01665-9.

[8] A.S. Perna, A. Astarita, D. Borrelli, A. Caraviello, F. Delloro, R. Della Gatta, P. Lomonaco, I. Papa, R. Sansone, A. Viscusi, Fused filament fabrication of ONYX-based composites coated with aluminum powders: a preliminary analysis on feasibility and characterization, (2021).

[9] M.C. Gryka, T.J. Comi, R.A. Forsyth, P.M. Hadley, S. Deb, R. Bhargava, Controlled dissolution of freeform 3D printed carbohydrate glass sca ff olds in hydrogels using a hydrophobic spray coating, Addit. Manuf. 26 (2019) 193–201. https://doi.org/10.1016/j.addma.2018.12.014.

[10] S. Slegers, M. Linzas, J. Drijkoningen, J. D'Haen, N. Reddy, W. Deferme, Surface roughness reduction of additive manufactured products by applying a functional coating using ultrasonic spray coating, Coatings. 7 (2017) 208. https://doi.org/10.3390/coatings7120208.

[11] R. Kumar, J.S. Chohan, R. Kumar, A. Yadav, Piyush, N. Singh, Hybrid fused filament fabrication for manufacturing of Al microfilm reinforced PLA structures, J. Brazilian Soc. Mech. Sci. Eng. 42 (2020) 1–13. https://doi.org/10.1007/s40430-020-02566-1.

[12] A. Viscusi, A. Astarita, D. Borrelli, A. Caraviello, L. Carrino, R. Della Gatta, V. Lopresto, I. Papa, A.S. Perna, R. Sansone, others, On the influence of manufacturing strategy of 3D-printed polymer substrates on cold spray deposition, (2021).

[13] A. Viscusi, R. Della Gatta, F. Delloro, I. Papa, A.S. Perna, A. Astarita, A novel manufacturing route for integrated 3D-printed composites and cold-sprayed metallic layer, Mater. Manuf. Process. 37 (2022) 568–581. https://doi.org/10.1080/10426914.2021.1942908.

[14] J. Zhu, J.L. Chen, R.K. Lade, W.J. Suszynski, L.F. Francis, Water-based coatings for 3D printed parts, J. Coatings Technol. Res. 12 (2015) 889–897. https://doi.org/10.1007/s11998-015-9710-3.

[15] M. Camović, A. Biščević, I. Brčić, K. Borčak, S. Bušatlić, N. Ćenanović, A. Dedović, A. Mulalić, M. Osmanlić, M. Sirbubalo, A. Tucak, E. Vranić, Coated 3D printed PLA microneedles as transdermal drug delivery systems, IFMBE Proc. 73 (2020) 735–742. https://doi.org/10.1007/978-3-030-17971-7_109.

[16] A. Milionis, C. Noyes, E. Loth, I.S. Bayer, Superhydrophobic 3D printed surfaces by dip-coating, Tech. Proc. 2014 NSTI Nanotechnol. Conf. Expo, NSTI-Nanotech 2014. 2 (2014) 157–160.

[17] K.M. Lee, H. Park, J. Kim, D.M. Chun, Fabrication of a superhydrophobic surface using a fused deposition modeling (FDM) 3D printer with poly lactic acid (PLA) filament and dip coating with silica nanoparticles, Appl. Surf. Sci. 467–468 (2019) 979–991. https://doi.org/10.1016/j.apsusc.2018.10.205.

[18] J. Žigon, M. Kariž, M. Pavlič, Surface finishing of 3D-printed polymers with selected coatings, Polymers (Basel). 12 (2020) 1–14. https://doi.org/10.3390/polym12122797.

[19] K.P. Akshay Kumar, K. Ghosh, O. Alduhaish, M. Pumera, Dip-coating of MXene and transition metal dichalcogenides on 3D-printed nanocarbon electrodes for the hydrogen evolution reaction, Electrochem. Commun. 122 (2021) 106890. https://doi.org/10.1016/j.elecom.2020.106890.

[20] C.T. Kao, C.C. Lin, Y.W. Chen, C.H. Yeh, H.Y. Fang, M.Y. Shie, Poly(dopamine) coating of 3D printed poly(lactic acid) scaffolds for bone tissue engineering, Mater. Sci. Eng. C. 56 (2015) 165–173. https://doi.org/10.1016/j.msec.2015.06.028.

[21] K.M. Lee, H. Park, J. Kim, D.M. Chun, Fabrication of a superhydrophobic surface using a fused deposition modeling (FDM) 3D printer with poly lactic acid (PLA) filament and dip coating with silica nanoparticles, Appl. Surf. Sci. 467–468 (2019) 979–991. https://doi.org/10.1016/j.apsusc.2018.10.205.

[22] N. Saleh, N. Hopkinson, R.J.M. Hague, S. Wise, Effects of electroplating on the mechanical properties of stereolithography and laser sintered parts, Rapid Prototyp. J. 10 (2004) 305–315. https://doi.org/10.1108/13552540410562340.

[23] K. Angel, H.H. Tsang, S.S. Bedair, G.L. Smith, N. Lazarus, Selective electroplating of 3D printed parts, Addit. Manuf. 20 (2018) 164–172.

[24] R. Bernasconi, G. Prioglio, M.C. Angeli, C.C.J. Alcantara, S. Sevim, S. Pané, P. Vena, L. Magagnin, Wet metallization of 3D printed microarchitectures: Application to the manufacturing of bioinspired microswimmers, J. Manuf. Process. 78 (2022) 11–21. https://doi.org/10.1016/j.jmapro.2022.03.057.

[25] C.G. Jones, B.E. Mills, R.K. Nishimoto, D.B. Robinson, Electroless deposition of palladium on macroscopic 3D-printed polymers with dense microlattice architectures for development of multifunctional composite materials, J. Electrochem. Soc. 164 (2017) D867.

[26] R. Bernasconi, C. Credi, M. Tironi, M. Levi, L. Magagnin, Electroless metallization of stereolithographic photocurable resins for 3D printing of functional microdevices, J. Electrochem. Soc. 164 (2017) B3059.

[27] E. Vaněčková, M. Bouša, R. Sokolová, P. Moreno-García, P. Broekmann, V. Shestivska, J. Rathouský, M. Gál, T. Sebechlebská, V. Kolivoška, Copper electroplating of 3D

printed composite electrodes, J. Electroanal. Chem. 858 (2020). https://doi.org/10.1016/j.jelechem.2019.113763.
[28] A.K. Sood, A. Equbal, Feasibility of FDM-electroplating process for EDM electrode fabrication, Mater. Today Proc. 28 (2019) 1154–1157. https://doi.org/10.1016/j.matpr.2020.01.099.
[29] A. Equbal, A.K. Sood, Investigations on metallization in FDM build ABS part using electroless deposition method, J. Manuf. Process. 19 (2015) 22–31.
[30] T. Juarez, A. Schroer, R. Schwaiger, A.M. Hodge, Evaluating sputter deposited metal coatings on 3D printed polymer micro-truss structures, Mater. \& Des. 140 (2018) 442–450.
[31] A. Afshar, D. Mihut, Enhancing durability of 3D printed polymer structures by metallization, J. Mater. Sci. Technol. 53 (2020) 185–191. https://doi.org/10.1016/j.jmst.2020.01.072.
[32] W. Tillmann, N.F. Lopes Dias, D. Stangier, L. Hagen, M. Schaper, F. Hengsbach, K.P. Hoyer, Tribo-mechanical properties and adhesion behavior of DLC coatings sputtered onto 36NiCrMo16 produced by selective laser melting, Surf. Coatings Technol. 394 (2020) 125748. https://doi.org/10.1016/j.surfcoat.2020.125748.
[33] R. Sputtering, D. Chioibasu, L. Duta, G. Popescu-pelin, N. Popa, N. Milodin, S. Iosub, L.M. Balescu, A.C. Galca, A.C. Popa, F.N. Oktar, G.E. Stan, A.C. Popescu, Animal Origin Bioactive Hydroxyapatite Thin Films Printed Cranial Implants, (2019).
[34] E. Chudinova, M. Surmeneva, A. Koptioug, P. Scoglund, R. Surmenev, Additive manufactured Ti6Al4V scaffolds with the RF–magnetron sputter deposited hydroxyapatite coating, J. Phys. Conf. Ser. 669 (2016). https://doi.org/10.1088/1742-6596/669/1/012004.
[35] P.P. Pokharna, M.K. Ghantasala, E.A. Rozhkova, 3D printed polylactic acid and acrylonitrile butadiene styrene fluidic structures for biological applications: Tailoring bio-material interface via surface modification, Mater. Today Commun. 27 (2021) 102348. https://doi.org/10.1016/j.mtcomm.2021.102348.
[36] A. Afshar, D. Mihut, Enhancing durability of 3D printed polymer structures by metallization, J. Mater. Sci. Technol. 53 (2020) 185–191. https://doi.org/10.1016/j.jmst.2020.01.072.

8 Post-Processing for Metal-Based Additive Manufacturing Techniques

Dhirenkumar Patel and Akash Pandey
The Maharaja Sayajirao University of Baroda,
Vadodara, Gujarat, India

CONTENTS

DOI: 10.1201/9781003276111-8

8.1 INTRODUCTION TO METAL ADDITIVE MANUFACTURING

Additive manufacturing is a type of manufacturing process which uses three-dimensional (3D) computer-aided-design (CAD) data to manufacture the part/object layer by layer in a printing manner, hence it is popularly known as 3D printing. Metal additive manufacturing is one of the most exciting research domain among researchers and technologists as it is reducing product development time and making additive manufacturing significant in mainstream manufacturing. Metal additive manufacturing processes are widely used to repair worn metallic high-value parts like a turbine blades, dies and molds, medical parts, aerospace parts, etc. [1]. Powder bed fusion, direct energy deposition, binder jetting, and metal extrusion are the metal additive manufacturing processes. Figure 8.1 shows the flow chart of metal additive manufacturing processes.

8.1.1 Brief on Metal AM

Powder bed fusion is a 3D manufacturing technique where a high energy source (laser beam/electron beam) is used to scan the selective area of each layer of powder to melt or sinter. The energy beam is reflected on the platform by a series of reflating mirrors and focused lenses. Reflective mirrors are used to scan each two-dimensional (2D) cross section of the metal powder layer. After each scan, the 2D cross-section platform is lowered down and a new layer of powder is spread by recoating the blade and

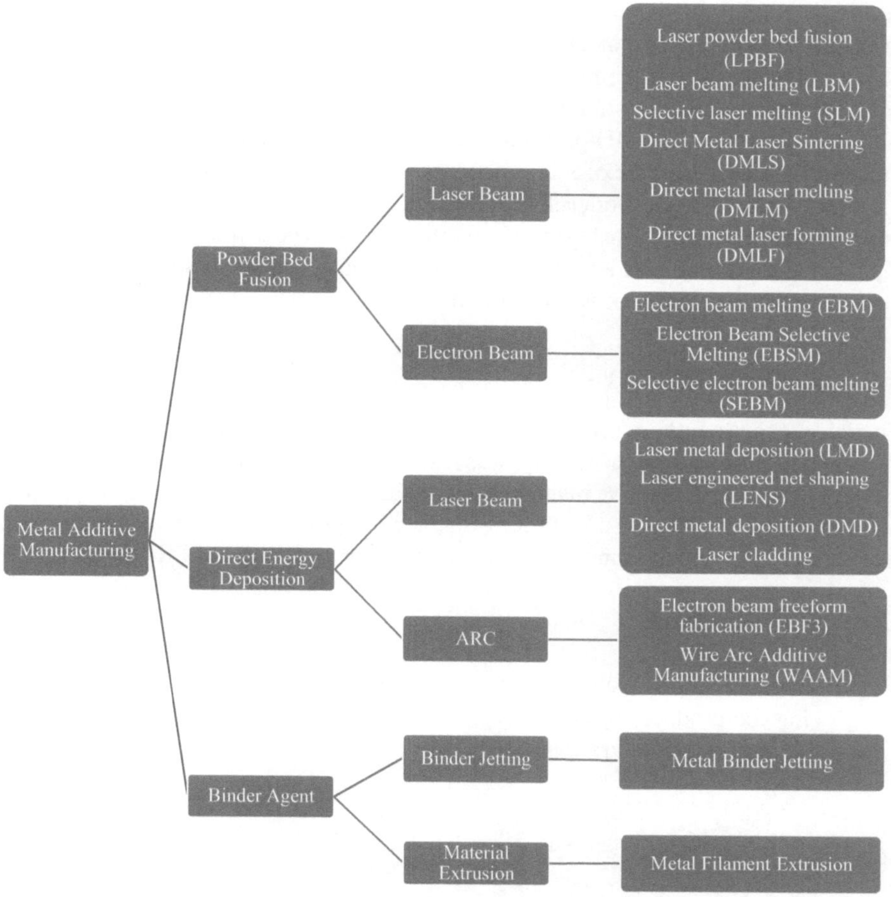

FIGURE 8.1 Metal additive manufacturing classification.

the next 2D cross-section is scanned and continued for the next one until the whole 3D part is produced. Selective laser sintering (SLS), selective laser melting (SLM), and electron beam melting (EBM) are powder bed fusion based additive manufacturing methods.

Direct energy deposition (DED) uses metal wire or powder as the raw material and laser, plasma and electron beam are used to generate heat to melt the raw material. DED is complex and based on the metal cladding principle. The machine nozzle is mounted on an A4 or five-axis robotic arm and adds the material to the substrate or existing part. Shielding gas is also fed from the nozzle to prevent oxidation during the printing process. Laser engineered net shaping (LENS), direct metal deposition (DMD), and wire arc additive manufacturing (WAAM) are direct energy deposition based additive manufacturing methods. Direct energy deposition is rapid and provides tremendous cost reduction possibilities for medium-to-large-size engineering components of moderate complexity.

Binder jetting for metal working is a similar method to powder bed fusion but instead of a heat source binder it is printed using an inkjet head to bind the metal powder together to form complex 3D objects. The printed part is often delicate and needs post-processing for strength improvement and densification.

Metal extrusion based 3D printing works in the same manner as fused deposition modeling 3D printing by extruding the metal blended thermoplastics material from the nozzle in *X* and *Y* axes for each layer deposition and the building platform lowered down for the next adjacent layer to continue for the whole printing process. After the complete printing part is sintered for removal of the plastic binder by controlled burning and infiltration process performed.

8.2 METHODOLOGY

This chapter was authored based on a review conducted according to the PRISMA systematically. The search activity was carried out between 14 August 2022–06 January 2023. Metal additive manufacturing or additive manufacturing or 3D printing and the keywords, defects, post-processing, need of post-processing, machining etc., were used as the search string. Figure 8.2 depicts the flow chart of PRISMA.

Consideration criteria

(1) The article must include metal additive manufacturing techniques for producing the product/object in any domain such as aerospace, medical, automobile, and manufacturing industry.

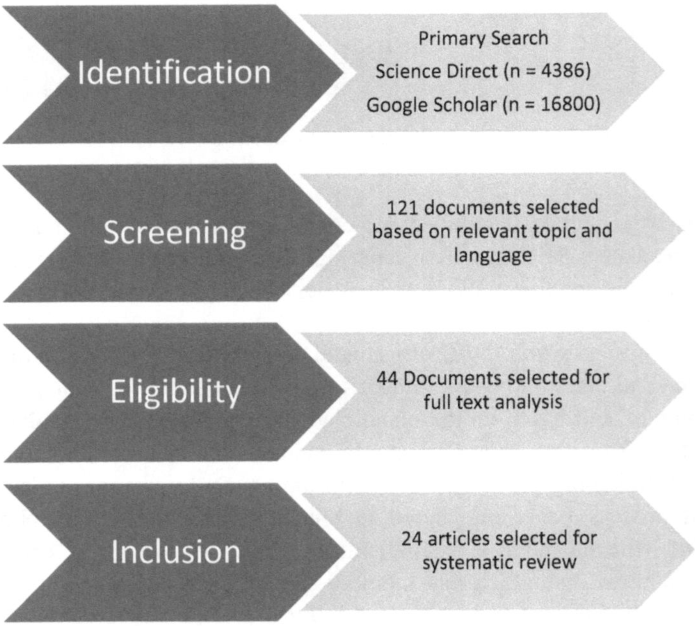

FIGURE 8.2 PRISMA flow chart of systematic review process.

(2) The article must focus on defects and post-processing of produced parts.
(3) The article includes only peer-reviewed research papers and experiment-based findings.
(4) The reviewed articles are in English only.

8.3 NEED OF POST-PROCESSING

Post-processing is needed in additive manufacturing due to imperfections produced during the process. Imperfections such as pores, cracks, non-isotropy, shape distortion, residual stress, and thermal stress were produced in the additively manufactured part. Defect formation is due to many reasons in metal additive manufacturing. Table 8.1 shows the summary of imperfections.

8.3.1 Porosity Formation

The formation of pores is a significant defect in additive manufacturing-based processes. Generally, two types of porosity formed in metal AM are keyhole pores and metallurgical pores. Keyhole pores are irregular in shape while metallurgical pores are regular and spherical. Keyhole pores are due to parameters like high speed or low power fails to fuse powder while metallurgical pores occurred during the low scan speed resulting in gas trapping and alloy evaporation. Pores have occurred during the process at the surface or sub-surface because the manufacturing of objects is done by layer-by-layer deposition. In electron beam-powder bed fusion, the pore is empty – literally a vacuum while in the case of laser powder bed fusion pores are gas-filled. Pores significantly make an impact on mechanical properties, corrosion resistance and fatigue behavior of the printed parts. Hence it is important to reduce or eliminate the pores in the printed part by applying the appropriate post-processing

TABLE 8.1
Imperfection of Metal Additive Manufacturing

Metal AM	Imperfections
Powder bed fusion	Poor surface roughness Residual stress Porosity and shrinkage Powder oxidation
Direct energy deposition	Dimensional inaccuracies, Rough surface finish Residual stress Lower hardness Oxidation
Metal binder jetting	Porosity, Rough surface finish
Metal material extrusion	Surface cracks

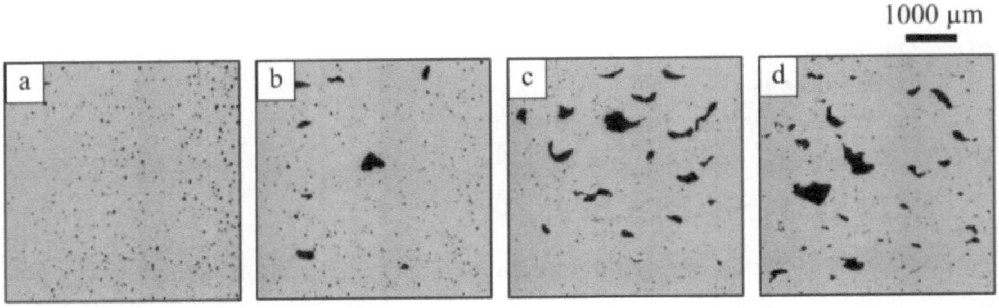

FIGURE 8.3 Formation of pores at different scanning speed: (a) 250 mm/s, (b) 500 mm/s, (c) 750 mm/s, and (d) 1000 mm/s. (source: Nesma T. 2014) [3].

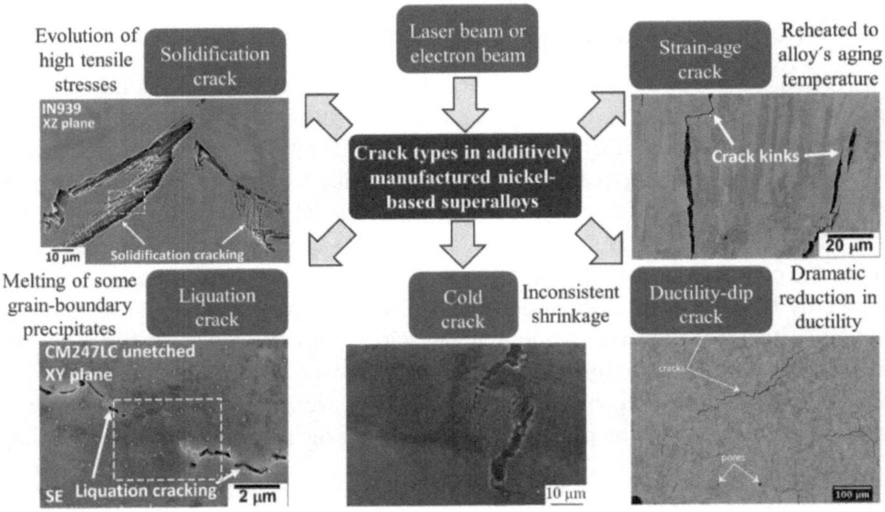

FIGURE 8.4 Different types of crack induced in metal additive manufacturing. (source: Qingsong Wei 2022) [5].

technique [2]. Optimum parameters can produce 99% dense parts while 1% of porosity can be reduced or eliminated by heat treatment. Figure 8.3 showing the porosity left to right: keyhole pores to metallurgical pores at different scan speeds keeping the same laser power.

8.3.2 Cracks

Crack formation is another defect, which plays a vital role in the mechanical performance of the printed part in additive manufacturing. Extensive use of the metal using additive manufacturing is limited because of cracks. Different types of metal material and alloys tend to produce different types of cracks. Figure 8.4 represents the different types of cracks such as solidification cracks, liquation cracks, strain-age cracks, ductility-dip cracks, and cold cracks. Crack elimination can be achieved by adding an appropriate element to the raw material to alter the composition for controlling the solidification [4].

8.3.3 Non-Isotropy

In metal additive manufacturing non-isotropy defines the imperfection of mechanical characteristics of additive manufactured parts. Thermal history and part printing orientation is playing major roles in the degree of non-isotropy [6]. The selection of appropriate laser power and scan speed with optimum scanning strategy can reduce the non-isotropy to some extent.

8.3.4 Surface Roughness

Additive manufacturing produces the part in a layer-by-layer manner hence it is producing a poor surface finish. Surface roughness can be improved by opting for the best input variables but at the expense of other significant properties.

8.3.5 Residual Stress

Residual stress induced in metal additive manufacturing is due to the instantaneous heating and cooling of the thermal cycle. Residual stress greatly affects the mechanical performance of the printed part, corrosion resistance, crack growth/fatigue, and dimensional stability [7]. Instantaneous heating causes the local expansion of the melt pool, which results in compression due to the surrounding unheated material after scanning rapid cooling took place, hence material contraction took place in the heated zone but it is partially restrained by the plastic formation during heating. Hence tensile stress formed in the melted zone, which is known as residual stress due to the thermal gradient mechanism [8].

8.4 ENHANCEMENT OF SURFACE QUALITY

Surface quality enhancement done by various post-processing methods like support removal, surface texture improvement, and aesthetics improvement.

8.4.1 Support Removal

Support removal is the most general type of post-processing applicable to all types of additive manufacturing processes, including metal additive manufacturing. There are two types of supports in metal additive manufacturing: (1) natural supports and (2) synthetic supports. Natural support is formed by surrounding materials during the build process while in the case of synthetic support, rigid structures in the form of thin walls, zig-zag lines, or wireframe skeleton designs and build as an attached part during the building process using the same raw materials. Supports are to be removed from the overhand parts and from the intricate internal features like holes, cavities, and internal overhanging flat or curvy surfaces.

8.4.1.1 Post-Processing of Natural Supports

In additive manufacturing, raw material act naturally as a support without design or instruction during the building process known as a natural support. In powder bed fusion and binder jetting, unfused powder material surrounding the build part acts as a natural support. In PBF of metal additive manufacturing, the part is being cooled by post build time by keeping it in the idle stage for a few minutes and surrounded by powder to eliminate the distortion to some extent. After the cool down stage is over, various processes are involved in the removal of support materials. In the case of PBF and binder jetting, the entire build volume along with the substrate taken to the dedicated breaking area for support removal manually. Various tools like abrasive brushes, compressed air, and glass bead blasting were used to remove powder stuck on the surface. Dentistry tools and some woodworking tools were used to remove partially sintered powder from the intricate feature or from the small internal holes or surface of the part. Nowadays, automated powder removal apparatus are available, which work based on vibration and vacuum to remove the support powder automatically.

8.4.1.2 Post-Processing of Synthetic Supports

Synthetic supports are rigid structures designed for the overhanging feature of the part to support, restrain, and attach the part to the build platform during the building process. While natural support acts naturally and it is in raw material form only. Synthetic supports tend to resist distortion in the case of PBF. In metal additive manufacturing synthetic supports are made of build material only. Synthetic supports are generated automatically using dedicated AM software and it depends on the part orientation and axis of the build. Orientation of support also plays an important role in the surface finish of desired part as it is going to keep small bumps or dents after support removal. Hence, appropriate orientation and support location are significant while designing the synthetic supports and removal of support is a bit tricky and needs extra attention. Removal of supports required pneumatic chisels, band saws, milling or wire cut EDM as metal supports cannot be removed by hand. Polishing and sanding are required to enhance the surface finish of the part after synthetic support removal. The example of the aerospace component with synthetic support is given in Figure 8.5.

Cabin bracket of Airbus A350 XWB passenger aircraft built by GE Additive's Laser CUSING R SLM additive manufacturing process [9].

8.4.2 Post-Processing for Surface Texture

Surface texture improvement is needed for aesthetic enhancement or functionality point of view. In metal material extrusion and direct energy deposition few surface texture imperfections, such as stair steps, powder adhesion, fill patterns, and marks of support removal are found on the printed part. Stair stepping can be reduced by employing minimum layer height at the cost of build time. Build orientation can reduce the powder adhesion to some extent. Post-processing, such as glass bead blasting can produce a matte surface texture and remove the sharp corners of the stair steps. While dry/wet sanding and polishing produce smooth or polished surfaces.

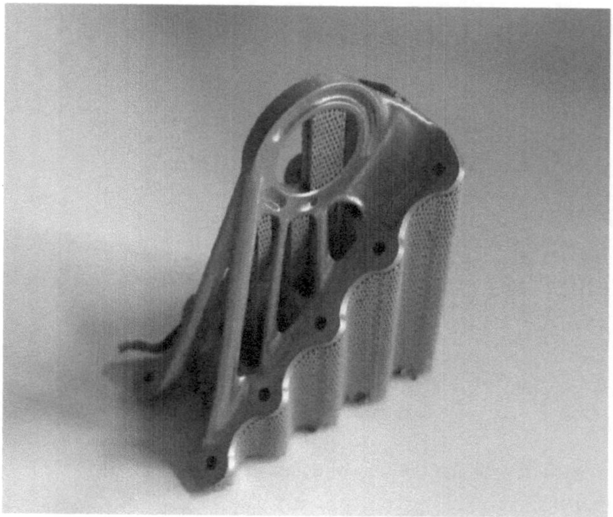

FIGURE 8.5 Aerospace component with synthetic support. (Adapted from *Metals* **2019**, *9*(6), 689) [9].

Paint and sealants were applied on the surface before or after the sanding for filling the porosity and smoothing the surface effectively. Automated techniques like tumbling and abrasive flow machining were used to machine the external and internal surfaces more precisely.

8.4.2.1 Laser Polishing

Laser polishing of additively manufactured parts/objects is carried out by remelting and solidification of a surface layer of the part to be super finished using a laser beam. Figure 8.6 shown the result of laser polishing carried out on knee joint part. Surface finish enhancement is a result of the smooth redistribution of molten material due to surface tension, which is generated by the thermal interaction of the laser with the surface of the part to be post-processed. Many types of laser polishing processes are employed in the post-processing of metal additive manufacturing, such as pulsed laser, continuous laser, a hybrid of pulsed and continuous laser and selective laser polishing techniques [10].

8.5 ENHANCEMENT OF DIMENSIONAL ACCURACY

Metal additive manufacturing produce a broad range of accuracy level starting from within micron to the unit millimeter. Accuracy depends on the size of the build volume and the speed of the building. Direct energy deposition accuracy depends on the speed of deposition. AM machine manipulator has some degree of accuracy based on the *XY* and *Z* axes motor control, which is dependent on the machine control and build architecture. Also in the case of thermal-based metal AM accuracy depends on the thermal gradient. Hence, the elimination of the dimensional imperfections is

FIGURE 8.6 Unprocessed and laser polished knee joint, produced by metal AM. (source: Fraunhofer ILT) [11].

achieved by implementing either compensation during design as pre-processing or by machining strategy as post-processing.

8.5.1 Machining Strategy as Post-Processing

Machining strategy means controlled material removal by selecting the appropriate machining process based on desired outcome and shape complexity of AM component. Selection of proper post-processing machining strategy for metal AM plays a vital role in achieving dimensional accuracy, surface quality, and clearance for functional requirement. Machining strategy are broadly classified into two types: (1) conventional machining, (2) advanced machining. Conventional machining is when material removal is because of shearing force and the tool is always more superior than the material to be removed while in the case of advanced machining material removal is due to shearing force but in the absence of mechanical cutting force and the tool is not superior compared to the material to be removed. Figure 8.7 depicts the various machining processes used in post-processing of metal AM parts.

8.5.2 Post-Processing by Conventional Machining

8.5.2.1 Grinding

Grinding is considered a prominent post-processing machining process to achieve the desired surface finish and dimensional accuracy in metal additive manufacturing. Superior materials having high hardness like carbide tips or tungsten carbide can be processed using metal AM and such a hard material can be processed using a CNC-based grinding process. Lapping and polishing along with rotary devices can be used to process rough surfaces but lapping and polishing can improve the surface but cannot contribute to dimensional accuracy like grinding. K. Philip Varghese, PHD.,

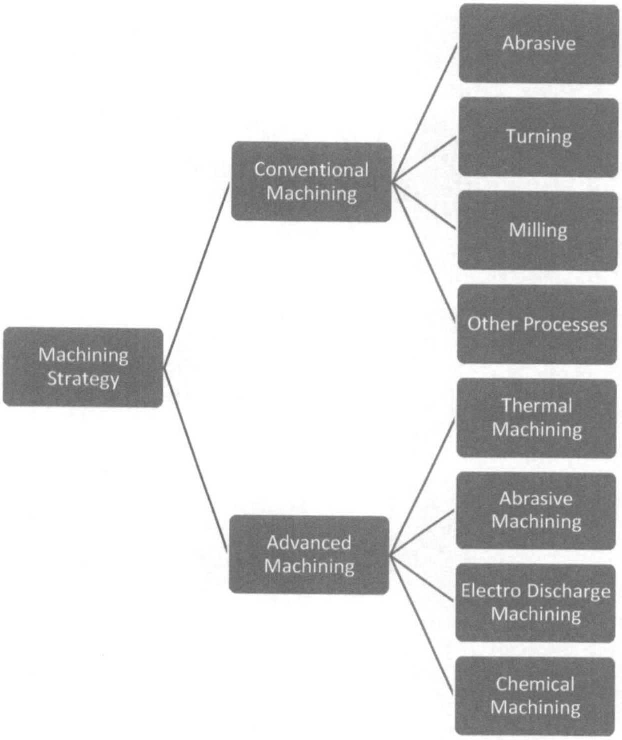

FIGURE 8.7 Classification of post-processing.

and John Hagan et al. of Saint-Gobain conducted experimentation of finish grinding for additively produced Inconel 718 specimens and the result reported improvement in surface finish. The average surface roughness of 0.21 microns from 3.3 microns was achieved after grinding in the case of parallel direction measurement and 0.5 microns in the case of perpendicular direction measurement. Figure 8.8 shows the improvement in the metal AM part. The grinding process is capable of producing a very smooth surface for scanning electron microscope (SEM) and electron back-scatter diffraction (EBSD) [12].

8.5.2.2 Turning

Turning is a metal removal process considered for cylindrical parts and it is most suited for additively manufactured metal parts for internal and external surfaces for better control over dimensional and geometrical accuracy. Multi-axis CNC turning is used for non-symmetrical shapes.

8.5.2.3 Milling

Milling of the workpiece is achieved by the interaction between the rotary tool and stationary workpiece. Multi-axis milling is used for complex parts easily by moving in 5–6 axes easily. Milling is producing a very superior surface and accurate surface

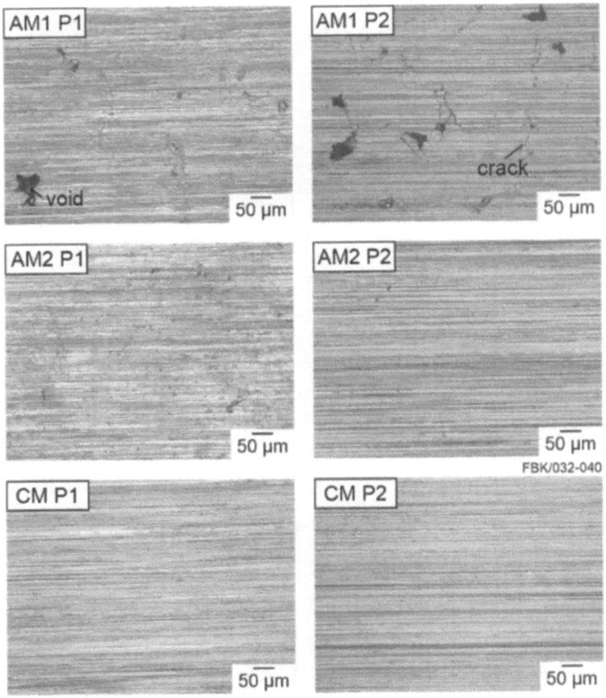

FIGURE 8.8 Surface after different grinding processes of additively manufactured AISI 316L (source: Benjamin Kirsch et. al. 2021) [13].

in comparison to turning and grinding. Milling is widely used post-processing for additively manufactured metal molds for injection mold and die-casting core [14].

8.5.2.4 Other Conventional Machining Processes

Post-processing of metal AM parts by hybrid machining, such as a combination of milling-turning can produce superior quality surfaces using turning and milling. A hybrid turning and milling machine can give 12 axes' movement opportunities as the turning head can rotate in *X*, *Y*, and *Z* directions and the milling head also can move in the other six axes.

8.5.3 Post-Processing by Advanced Machining

Advanced machining involved the thermal, abrasive, plasma, and chemical means to remove the material.

8.5.3.1 Thermal Machining

In thermal machining, material removal is achieved by heating, melting, metal vaporization, and chemical degradation at high temperatures. A thermal energy source such as a laser, plasma arc, electron beam, and ion beam is used for post-processing.

8.5.3.1.1 *Laser Machining*

Laser machining post-processing can be utilized to improve the dimensional accuracy of metal AM parts by surface enhancement. The effectiveness of laser machining depend on the thermal conductivity and absorptivity of the material. Laser machining is able to machine any metal alloy material irrespective of hardness. The selection of laser machining is a bit calculative as it may induce defects while machining and it can adversely affect the fatigue property of additively manufactured parts.

8.5.3.1.2 *Plasma Arc Machining*

Plasma arc generation occurred beyond 3000 °C temperature by electrons and ionized gas. Plasma arc has high thermal energy, which can be utilized for any metal material. In this post-process, the part is heated up to melting temperature and plasma pressure is applied to remove the material in a controlled manner. Cutting, grooving, and face turning can be done by plasma arc machining effectively. Plasma arc machining tends to produce residual stresses hence selection for post-processing needs to be careful consideration.

8.5.3.1.3 *Electron Beam Machining (EBM)*

Electron beam machining works in a vacuum with a focus area of 25 microns at high energy density. The energy is absorbed by the additively manufactured metal part and evaporation takes place hence material removal happened. Electron beam machining produces very good surface quality. For gold and carbon surface finish achieved up to 5 μm and for titanium, it is up to 10 μm. Electron beam machining is able to produce holes of 25 μm with a depth of 10 mm length. Electron beam machining is suitable for hard brittle and thin cross-sections. Electron beam machining is expensive and produces residual stress in the part.

8.5.3.1.4 *Ion Beam Machining (IBM)*

Ion beam machining functions in the same way as electron beam machining. In IBM plasma source and electromagnetic coils were used to ionize the gas atoms using a heated tungsten cathode, which emits electrons when a high potential difference is applied, electromagnetic coils were used to give a path to flow electrons into the helical spiral pattern to travel more length. Hence ionization reaches the peak and generates high energy, which evaporates the metal material from the workpiece, which results in material removal in a highly controlled way. IBM is highly used for machining tiny components, such as 100 nm with surface quality reported to be 1 μm. IBM operates at a low temperature compared to the other thermal-based processes and has no chemical reaction hence easy to control but this process required an expensive apparatus [15].

8.5.3.2 Abrasive Machining Processes

Abrasive machining processes are significant for metal additive manufacturing, water jet machining (WJM), abrasive water jet machining (AWJM), abrasive flow machining (AFM), abrasive jet machining (AJM), abrasive barrel machining (ABM), and ultrasonic machining (USM) are used for post-processing.

8.5.3.2.1 *Water Jet Machining (WJM)*

Water jet machining uses high-pressurized water from normal pressure to 3800 bars without pressure fluctuations. This high-pressure water is concentrated by a hard nozzle at the speed of 900 m/s or even more. Water jet machining performs cutting but in some cases, it is used for surface machining by increasing the nozzle diameter. Water jet machining is used for removing the sharp edges/burrs from the organic contours in the case of powder bed fusion MAM [16].

8.5.3.2.2 *Abrasive Water Jet Machining (AWJM)*

AWJM has the same working principle as WJM. But the only key difference between the process is the use of abrasive particles in the fluid flow. An abrasive feeder is used to provide abrasive particles into the water stream without fluctuations and a mixture of water and abrasive particle increases the performance of cutting any hard metal material. The advantages of using AWJM for the metal AM process are low-temperature machining with intricate features, clean surfaces, and it being good for complex shapes. AWJM is highly recommended for thermal-based metal additive manufacturing where residual stress is a matter of concern as cyclic heating and cooling induce the residual stress. Another process called abrasive jet machining works in a similar way as AWJM but it is used air as a medium to carry abrasive particles.

8.5.3.2.3 *Abrasive Flow Machining (AFM)*

Abrasive flow machining employs the abrasive-laden fluid flow back and forth through the internal part of the surface to be post-process. AFM effectively post-process the intricate features of the additively manufactured products, such as lattice structure and it is the most appropriate selection for internal surface features like holes, slots, and cavities, which are normally difficult or not possible to post-process using other conventional polishing and grinding process. AFM is successfully used for metal additively manufactured components. AFM is suitable for the small surface area, specifically internal surfaces.

8.5.3.2.4 *Abrasive Barrel Machining (ABM)*

In ABM, workpieces are put into the rotating barrel along with a mixture of fluid and abrasive particles. As the barrel rotates, the rotation movement of the barrel makes the interaction between the abrasive particles and the workpiece result in material removal. Process variables such as speed of rotation, size, and quantity of abrasive particles and geometry of the workpiece significantly affect the metal removal rate (MRR). The use of water in the process makes the low operating temperature, which contributes to no residual stress generation. Therefore, ABM is highly suitable for metal AM parts for machining complex shapes and the only demerit is the process does not have any control over the dimensional accuracy uniformly.

8.5.3.2.5 *Ultrasonic Machining (USM)*

In ultrasonic machining, abrasive grit particle indents on the surface to be post-processed by the tool and the tool vibrates linearly. Abrasive particles having a hardness of greater than 40 HRc are used in post-processing. The vibration is produced by an ultrasonic transducer with a high frequency of 20 KHz. USM is capable of post-processing any hard or brittle material easily. Surface finish can be achieved in the range of 7–25 microns. USM can be employed for the machining of thin walls produced by powder bed fusion-based metal AM. USM produced less residual stress as the operating temperature is low during the material removal.

8.5.3.3 Electrical Discharge Machining (EDM)

In electro discharge machining, the potential difference is applied to the cathode electrode and a component to be post-processed. The component is filled with dielectric fluid. Sparks generated between the cathode electrode tool and component cause the instantaneous temperature generation in the range of 12000–14000 °C. Such high-temperature melts and evaporates the metal from the surface and material removal takes place. EDM is a non-tactile process, which exerts less or no mechanical force during the process, suitable for the post-processing of metal AM parts. Its peculiarity to machine thin walls is unable in the conventional machining process. EDM produces a good surface finish without leaving any burrs without distortion. Wire EDM is generally utilized for additively manufactured part removal from the build plate. Wire cut EDM cut the component without any mechanical force with the kerf width of 0.2 mm, the same as the diameter of the wire. Hence, EDM is a widely accepted post-processing technique in metal additive manufacturing [17].

8.5.3.4 Chemical Machining Processes

In chemical machining, material removal is done by corrosive solutions in a temperature-regulated etching bath. Masking material is used to protect the area or surface for no machining. Chemical machining is highly suitable for complex geometry, which can be distorted/damaged by cutting forces in conventional machining. ChM provides uniform material removal from the surface of the workpiece, which makes ChM highly suitable for the post-processing of additively manufactured jewelry and parts of the aerospace segment. ChM produces gas from the corrosive solution and the material removal rate is less. The automated post-processing machine for metal AM was developed by an Australian company named Hirtenberger. The automated process used a combination of hydrodynamic flow, electrochemical pulsing, and particle-assisted surface cleaning and material removal. In this automated process support removal and rough or fine finish are achieved effectively. This automated system is suitable for most of the metals and their alloy irrespective of the material's hardness with a surface quality of 0.5 microns, which is highly recommended in critical applications.

8.6 POST-PROCESSING USING NON-THERMAL TECHNIQUES

Non-thermal post-processing techniques are used for improvement in the mechanical performance of metal AM products/components.

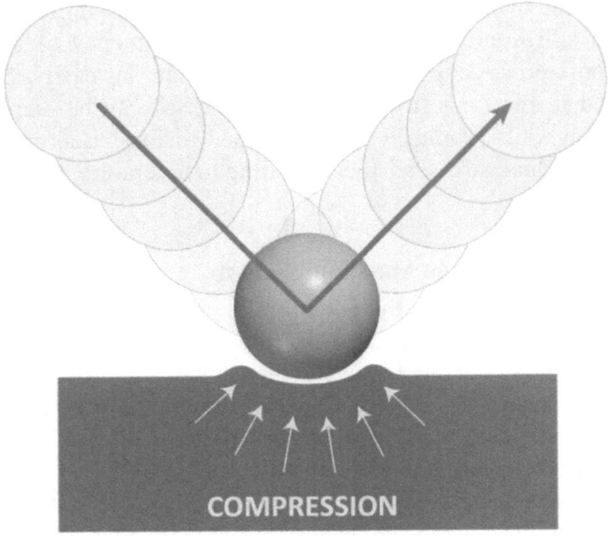

FIGURE 8.9 Principle of shot peening.

8.6.1 Shot Peening

Shot peening is the process of inducing compressive residual stress by bombarding the super-hard spherical media on the surface of the workpiece in a controlled manner. The bombarding media commonly glass, ceramic, or hard steel powder with higher hardness compared to the material to be processed. In the shot peening, spherical particle indents in the surface like a peening hammer and create a dimple on the surface. Figure 8.9 depicts the working principle of shot peening.

In thermal-based additive manufacturing such as metal material extrusion, powder bed fusion and direct energy deposition, cyclic heating and cooling induce high residual stress within the workpiece surface. Shot peening induces compressive residual stress, which proved beneficial. Improvement in mechanical properties such as fatigue resistance, fretting, corrosion cracking, and reduction in internal stress is significantly achieved with the help of shot peening in metal additive manufacturing. Metal AM produces intricate and lattice shapes in the component, which can be post-processed easily and effectively by small blasting media of shot peening. Some metal AM powder can be used as a blast particle in shot peening if MAM powder is harder in nature and further it can be used/recycled in building the part. Multi-axis shot peening is used in the post-processing of additively manufactured gears and turbine blades made of metal material.

8.6.2 Cold Isostatic Pressing (CIP)

Cold isostatic pressing is a post-processing method of compacting additively manufactured metal parts into solid homogeneous parts before machining or sintering. Cold isostatic pressing is also known as hydrostatic pressing. The working pressure in cold isostatic pressing is 15000 to 60000 psi at the ambient temperature of around 93 °C. wet and dry types of CIP used in post-processing. In wet CIP, the workpiece is surrounded by the fluid to exert uniform pressure to enhance the uniformity of the part. For simple geometric parts, pressure on the workpiece is applied by means of a bag or mold. But for complex or near-net-shape components, the component surface directly comes into contact with the fluid. CIP is most appropriate for the consolidation of metal AM parts. Hence CIP is an ideal post-processing method.

8.7 POST-PROCESSING USING THERMAL TECHNIQUES

Thermal post-processing techniques are classified into three categories based on applied pressure as shown in Figure 8.10. Heat treatment is the most prominent way to achieve the surface quality, mechanical properties, and material integrity of metal AM parts by eliminating porosity to form a homogeneous microstructure.

8.7.1 Heat Treatment in Ambient Pressure

Thermal post-processing is done on AM parts to enhance mechanical performance. In powder bed fusion and direct energy deposition, heat treatment is the primary post-processing to improve the microstructure and relieve residual stress. New specific heat treatment methods have been developed for retaining the fine-grained microstructure within the additively manufactured part while relieving the stress and improvement. It is necessary to create a custom heat treatment technique when the AM-built part has a non-equilibrium microstructure that doesn't respond as expected to traditional heat treatment methods. Many new post-processing techniques were designed and developed to get dense usable metal parts from the green metal part. Binder jetting is the key AM process ,which uses infiltration and sintering in the furnace. Figure 8.11 shows the steps involved in the infiltration post-processing of the binder jetting metal AM process.

An oil and gas stator printed using binder jetting from the stainless steel powder material and post-process by bronze infiltration to improve the surface finish.

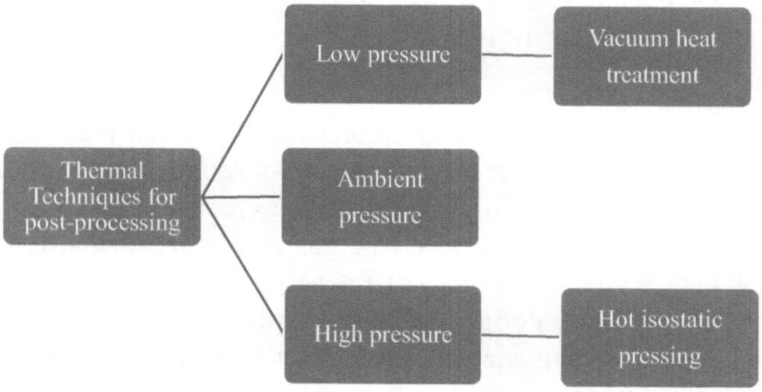

FIGURE 8.10 Classification of thermal post-processing technique.

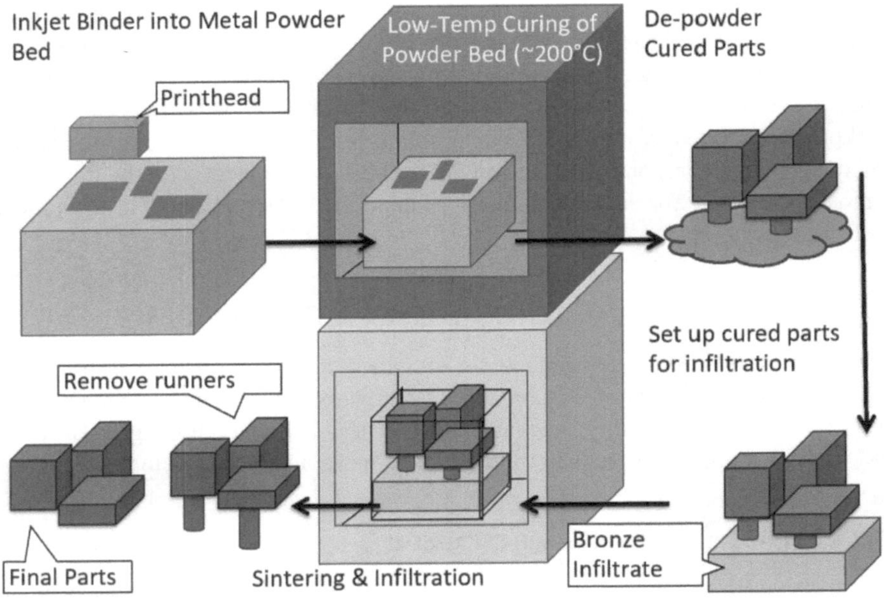

FIGURE 8.11 Schematic flow of sintering and infiltration process (source: Amir Mostafaei 2021) [18].

FIGURE 8.12 (A–C) Post-processed additive manufactured part by sintering and Al infiltration (source: Corson L 2019) [19].

Figure 8.12 shows the typical surface finish produced by binder jetting. Dimensional accuracy and shrinkage control are difficult to control and require the optimization of process parameters for such a kind of post-processing.

Selective laser sintering is used to create complex tools for the EDM cathode. Additive manufacturing of the green part for polymer-coated ZrB2 metal powder. Followed by debinding and sintering of the printed part and in the last step infiltration done by liquid copper. By this method, the EDM tool is manufactured with a more homogeneous structure compared to any post-processing technique.

8.7.2 Heat Treatment in Low Pressure

In metal additive manufacturing oxidation is the biggest challenge to control or manage. Hence heat treatment should be carried out in a vacuum or in a gas chamber using inert gas. The vacuum system of post-processing is more expensive in comparison to ambient temperature post-processing heat treatment. Furnaces operate at a temperature ranging 800–3000 °C uniformly in vacuum pressure with negligible contamination of other gases, such as carbon dioxide, oxygen, and other gases. In low-pressure heat treatment post-processing higher purity in the printed metal part is achieved by removing low-temperature impurities using a vacuum pumping system and the treated part reach the non-metallurgical level after the post-processing and is often rapidly cooled by argon inert gas [20]. Figure 8.13 shows the microstructure of annealed specimen of CoCrFeNi produced by cold-sprayed additive manufacturing. The annealing post-process reduces the deformation by recrystallization and grain size reformation, which enhances the mechanical performance [21].

8.7.3 Heat Treatment in High Pressure

8.7.3.1 Hot Isostatic Pressing

Hot isostatic pressing is a high-temperature high-pressure post-processing technique carried out at a very high-pressure range from 1500 to 30000 psi. Applied pressure is uniform from all sides of the object. HIP is one of the most prominent post-processing techniques as it gives an improvement in porosity by densification of powder metal parts, fatigue resistance, high impact and wear resistance, ductility enhancement and reduce machining work. Direct energy deposition and powder bed use different alloys for building the components. HIP can provide the best diffusion bond, which increases the bonding strength among such dissimilar metals [22]. Process parameters such as temperature and pressure are decided based on the properties of the material to be processed. HIP eliminate the porosity from the part made by metal-based extrusion effectively.

The hot isostatic process is used with powder bed fusion in a hybrid manner to produce parts by sintering followed by HIP. In this hybrid process outside of the perimeter contour is built by selective laser sintering and inside the hollow space metal powder is filled and then this part is processed by HIP to form a final dense component. This approach has no adverse effect on the property and it takes less time in comparison to the normal way of doing metal AM. This approach is effectively used for building complex 3D parts of Inconel alloy and titanium alloy for the medical and aerospace industry [23]. The cost of the process and dissolution of gas into the printed part is considered a demerit of this post-process. Karami et al. [24] carried out experimentation of HIP on AM-ed Ti6Al4V lattice structures, which is presented in Figure 8.14.

8.7.3.2 Laser Shock Peening

Peening is a process that plastically compresses material normal to a surface resulting in transverse (Poisson) expansion. In laser shock peening pressure pulses are used

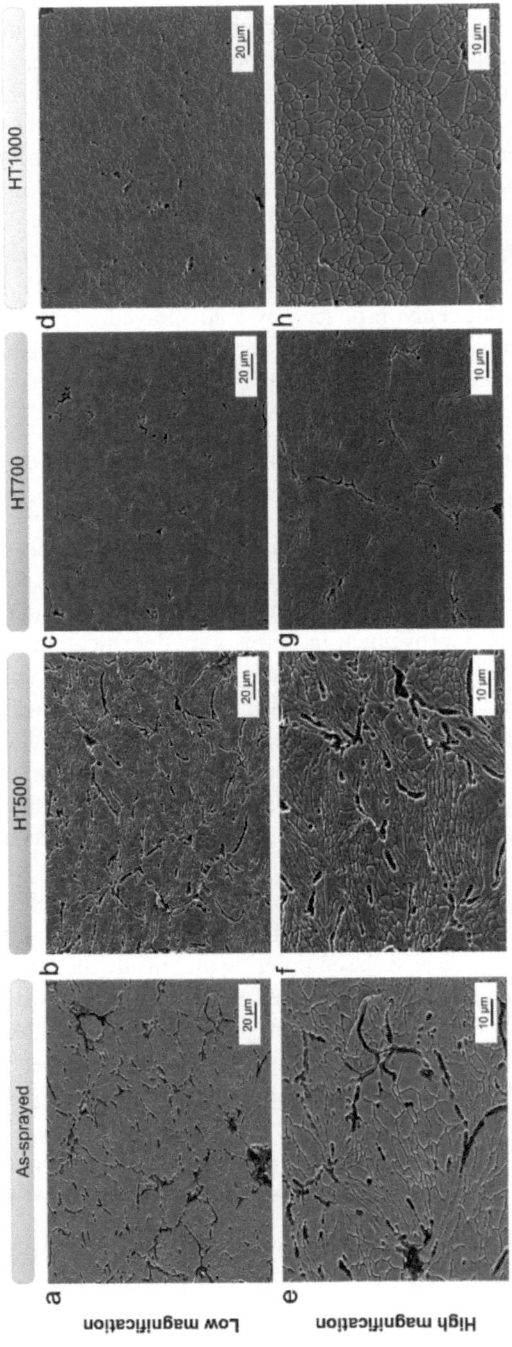

FIGURE 8.13 Microstructure of additively manufactured CoCrFeNi specimen at different annealing temperatures: (a, e) as-sprayed, (b, f) 500 °C, (c, g) 700 °C, and (d, h) 1000 °C [21].

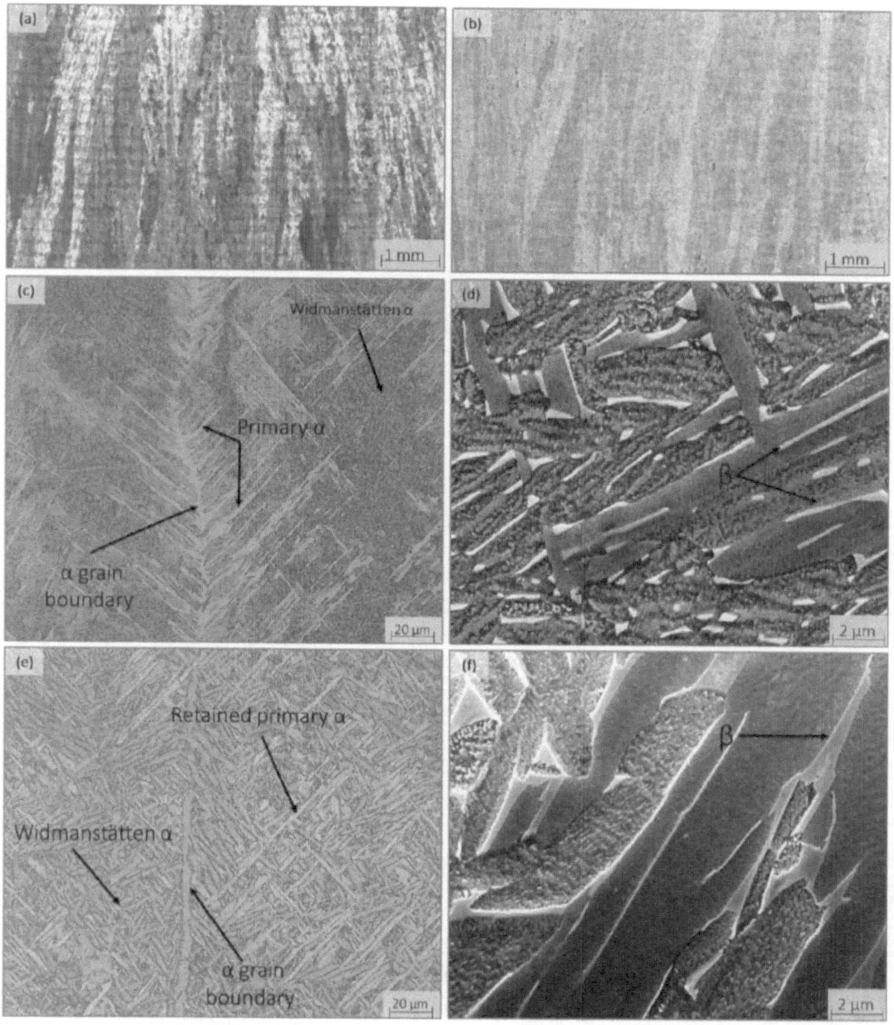

FIGURE 8.14 Ti6Al4V EBM printing (a) as-built macrostructure, (b) macrostructure after HIP, (c, d) as built microstructure, and (e, f) HIP microstructure. (Adapted from *Metals* **2022**, *12*(1), 77) [25].

to generate local plastic deformation by using short intensive laser pulses for 30 ns. Water is used to generate the pressure efficiently. A schematic of laser shock peening is shown in Figure 8.15. LSP substitutes the non-desirable tensile stresses formed in components during the additive manufacturing process with positive compressive stresses that improve part material characteristics. LSP decreases the voids near the surface formed by metal additive manufacturing in the same way that it eliminates voids in powdered metal components manufactured using more conventional processes [26].

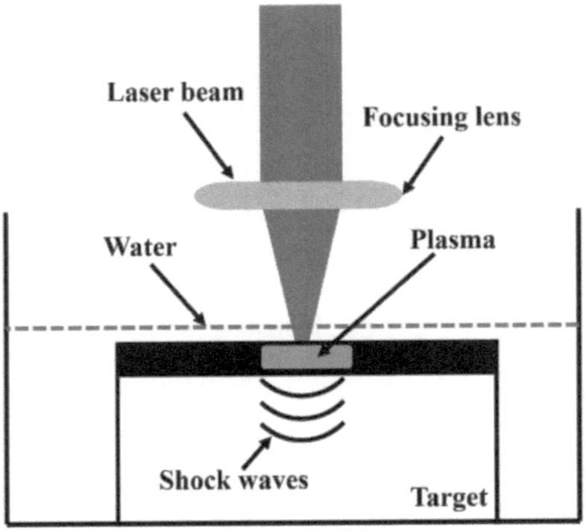

FIGURE 8.15 Laser shock peening. (Adapted from *Metals* **2022**, *12*(1), 77) [25].

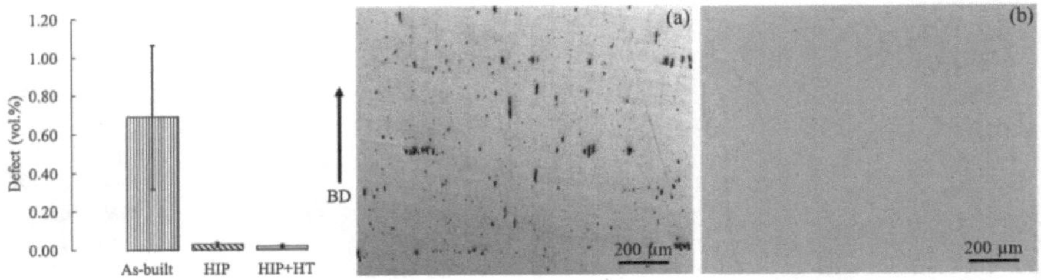

FIGURE 8.16 (a) Defect analysis, (b) optical micrograph of defects in the as-built material. (Adapted from *Metals* **2022**, *12*(1), 77) [25].

The fatigue life of metal additive manufactured components can be achieved by laser shock peening. The fatigue lifecycle of jet engine fans, compressor blades, aircraft structure, and nuclear fuel canisters are improved aggressively by laser shock peening [27, 28]. LSP is used to enhance the surface quality of additively manufactured maraging steel [29]. LSP is also used to get the desired curvature and stretching to the thick sections of aircraft wing panels providing precise aerodynamic form.

Goel et al. [30] investigated the effects of two different post-processing treatments: (a) HIP and (b) HIP + heat treatment on Inconel 718 alloy manufactured by electron beam melting (EBM) where heat treatment gives an improvement in the part microstructure as shown in Figure 8.16. Laser peening might be a significant post-processing in metal additive manufacturing as metal additive manufacturing (MAM), began as a popular mainstream manufacturing technology. Today, Airbus has over 1000 3D printed parts on its A-350 XWB planes, and Lockheed Martin and Boeing are also using the technology extensively. LSP could eliminate the need for HIP or heat treatment of the entire parts and the accompanied challenges.

8.8 CONCLUSIONS

This chapter discussed the need and importance of post-processing in metal additive manufacturing as most of the metal additive manufactured parts require post-processing before being put into the final application. Metal additive manufacturing requires the know-how of post-processing along with the merits and demerits of each additive manufacturing for its better and more extensive utilization in mainstream manufacturing. Also discussed are the imperfections, such as pores, cracks, anisotropy, residual stresses, and poor surface finish induced in metal additive manufacturing during the build process. Selection of post-processing technique based on form, fit, and function to be considered and explained. Removal of support material, enhancement of surface texture, and aesthetic enhancement are usually done for the desired shape. For the desired fit, and accuracy correction done by conventional machining processes like, milling, turning, grinding, etc., for the desired function, various thermal and non-thermal processes do property enhancement to improve mechanical performance. The concluding statement is "Post-processing has a significant impact on parts built by metal additive manufacturing as it can convert crack-prone parts, which can break by hand easily to robust parts for critical application."

ACKNOWLEDGEMENTS

We would like to acknowledge and give our regards to Dr. Eujin Pei and Dr. Gurminder for giving the opportunity to explore and write. We would like to thank our family members for constant support and motivation.

REFERENCES

[1] Yasa, E., Poyraz, O., Cizioglu, N., & Pilatin, S. (2015). Repair and manufacturing of high performance tools by additive manufacturing. *Notes*, *148*(135), 107.

[2] Al-Maharma, A. Y., Patil, S. P., & Markert, B. (2020). Effects of porosity on the mechanical properties of additively manufactured components: a critical review. *Materials Research Express*, *7*(12), 122001.

[3] Aboulkhair, N. T., Everitt, N. M., Ashcroft, I., & Tuck, C. (2014). Reducing porosity in AlSi10Mg parts processed by selective laser melting. *Additive manufacturing*, *1*, 77–86.

[4] Aversa, A., Lorusso, M., Cattano, G., Manfredi, D., Calignano, F., Ambrosio, E. P., ... & Pavese, M. (2017). A study of the microstructure and the mechanical properties of an AlSiNi alloy produced via selective laser melting. *Journal of Alloys and Compounds*, *695*, 1470–1478.

[5] Wei, Q., Xie, Y., Teng, Q., Shen, M., Sun, S., & Cai, C. (2022). Crack types, mechanisms, and suppression methods during high-energy beam additive manufacturing of nickel-based superalloys: A review. *Chinese Journal of Mechanical Engineering: Additive Manufacturing Frontiers*, 100055.

[6] Kempen, K., Thijs, L., Van Humbeeck, J., & Kruth, J. P. (2012). Mechanical properties of AlSi10Mg produced by selective laser melting. *Physics Procedia*, *39*, 439–446.

[7] Chen, S. G., Gao, H. J., Wu, Q., Gao, Z. H., & Zhou, X. (2022). Review on residual stresses in metal additive manufacturing: formation mechanisms, parameter

dependencies, prediction and control approaches. *Journal of Materials Research and Technology*.

[8] Mercelis, P., & Kruth, J. P. (2006). Residual stresses in selective laser sintering and selective laser melting. *Rapid Prototyping Journal*.

[9] Childerhouse, T., & Jackson, M. (2019). Near net shape manufacture of titanium alloy components from powder and wire: A review of state-of-the-art process routes. *Metals*, *9*(6), 689.

[10] Krishnan, A., & Fang, F. (2019). Review on mechanism and process of surface polishing using lasers. *Frontiers of Mechanical Engineering*, *14*, 299–319.

[11] Laser Polishing of Metals–Brochure–Fraunhofer ILT [www.ilt.fraunhofer.de/content/dam/ilt/en/documents/brochure/b-laser-polishing-of-metals.pdf (n.d) Fraunhofer Institute for Laser Technology ILT.

[12] www.mapyourshow.com/mys_shared/imts18/handouts/IMTS38KPV_IMTS2018.pdf

[13] Kirsch, B., Hotz, H., Hartig, J., Greco, S., Zimmermann, M., & Aurich, J. C. (2021). Pendulum and creep feed grinding of additively manufactured AISI 316L. *Procedia CIRP*, *101*, 166–169.

[14] Stucker, B., & Qu, X. (2003). A finish machining strategy for rapid manufactured parts and tools. *Rapid Prototyping Journal*, *9*(4), 194–200.

[15] Miller, B. D., Gan, J., Madden, J., Jue, J. F., Robinson, A., & Keiser Jr, D. D. (2012). Advantages and disadvantages of using a focused ion beam to prepare TEM samples from irradiated U–10Mo monolithic nuclear fuel. *Journal of Nuclear Materials*, *424*(1–3), 38–42.

[16] Phokane, T., Gupta, K., & Gupta, M. K. (2019). Near net shape manufacturing of miniature spur brass gears by abrasive water jet machining. *Near Net Shape Manufacturing Processes*, 143–158.

[17] Dodun, O., Gonçalves-Coelho, A. M., Slătineanu, L., & Nagîţ, G. (2009). Using wire electrical discharge machining for improved corner cutting accuracy of thin parts. *The International Journal of Advanced Manufacturing Technology*, *41*(9–10), 858.

[18] Mostafaei, A., Elliott, A. M., Barnes, J. E., Li, F., Tan, W., Cramer, C. L., … & Chmielus, M. (2021). Binder jet 3D printing—Process parameters, materials, properties, modeling, and challenges. *Progress in Materials Science*, *119*, 100707.

[19] Cramer, C. L., Elliott, A. M., Kiggans, J. O., Haberl, B., & Anderson, D. C. (2019). Processing of complex-shaped collimators made via binder jet additive manufacturing of B4C and pressureless melt infiltration of Al. *Materials & Design*, *180*, 107956.

[20] Khorasani, A., Gibson, I., Goldberg, M., & Littlefair, G. (2017). On the role of different annealing heat treatments on mechanical properties and microstructure of selective laser melted and conventional wrought Ti-6Al-4V. *Rapid Prototyping Journal*, *23*(2), 295–304.

[21] Fan, N., Rafferty, A., Lupoi, R., Li, W., Xie, Y., & Yin, S. (2023). Microstructure evolution and mechanical behavior of additively manufactured CoCrFeNi high-entropy alloy fabricated via cold spraying and post-annealing. *Materials Science and Engineering: A*, 144748.

[22] Shao, S., Mahtabi, M. J., Shamsaei, N., & Thompson, S. M. (2017). Solubility of argon in laser additive manufactured α-titanium under hot isostatic pressing condition. *Computational Materials Science*, *131*, 209–219.

[23] Das, S., Wohlert, M., Beaman, J. J., & Bourell, D. L. (1999). Processing of titanium net shapes by SLS/HIP. *Materials & Design*, *20*(2–3), 115–121.

[24] Karami, K., Blok, A., Weber, L., Ahmadi, S. M., Petrov, R., Nikolic, K., … & Popovich, V. A. (2020). Continuous and pulsed selective laser melting of Ti6Al4V

lattice structures: Effect of post-processing on microstructural anisotropy and fatigue behaviour. *Additive Manufacturing*, *36*, 101433.

[25] Mahmood, M. A., Chioibasu, D., Ur Rehman, A., Mihai, S., & Popescu, A. C. (2022). Post-processing techniques to enhance the quality of metallic parts produced by additive manufacturing. *Metals*, *12*(1), 77.

[26] Ding, K., & Ye, L. (2006). *Laser shock peening: performance and process simulation*. Woodhead Publishing.

[27] Anonymous, Boeing Awards MIC Laser Peening Contract to Form 747–8 Wing Sections, WrightCurtiss-, 2008.

[28] Anonymous, Laser peening by CWST contributes to successful nuclear canister storage program, Shot Peener Magaz. 32 (2) (2018).

[29] Raja, K., Balram, T. P., & Naiju, C. D. (2018). *Study of surface integrity and effect of laser peening on maraging steel produced by lasercusing technique* (No. 2018-28-0094). SAE Technical Paper.

[30] Goel, S., Sittiho, A., Charit, I., Klement, U., & Joshi, S. (2019). Effect of post-treatments under hot isostatic pressure on microstructural characteristics of EBM-built Alloy 718. *Additive Manufacturing*, *28*, 727–737.

9 Post-Processing of Additive Manufactured Components Through Abrasive Flow Machining Process

Sahil Sharma[1], *Tarlochan Singh*[2], *and Akshay Dvivedi*[1]
[1]Mechanical and Industrial Engineering Department, Indian Institute of Technology, Roorkee
[2]Product and Industrial Design Department, Lovely Professional University, Phagwara, Punjab, India

CONTENTS

9.1 INTRODUCTION

The continued development of materials and new machining methods are essential to meet industry requirements [1,2]. Additive manufacturing (AM) is a rapidly growing field, and metal-based additive manufacturing is at the forefront of this growth [3]. Metal-based additive manufacturing is already being used to create a wide variety of products, from medical implants to aerospace components [4]. Traditional metal manufacturing methods often require extensive tooling and complex machining operations, which can be both time-consuming and expensive. On the other hand, AM can produce metal parts with very little waste and without

DOI: 10.1201/9781003276111-9

the need for tooling. This makes it an ideal technology for producing metal parts quickly and affordably, whether for prototypes or production runs. The classification and characterization of various additive manufacturing processes are depicted in Figure 9.1.

Metal-based additive manufacturing (AM) process starts with creating a three-dimensional (3D) model of the desired object through a 3D scanning software or computer-aided design (CAD). Subsequently, the model is sliced into thin layers, and each layer is printed one by one at a time. The various steps associated with the process are depicted in Figure 9.2.

The AM processes can fabricate metallic components in a single step without needing special tools and fixtures. However, the fabricated components are not suitable to use for intended applications. This is because the surface of AM fabricated metallic components are often rough and requires finishing to achieve the desired level of smoothness. Therefore, metallic-based additive components are often post-processed in order to improve their properties and performance. Abrasive flow machining (AFM) is one widely used in the non-conventional post-processing process in which abrasive-laden slurry is forced through under pressure to improve the surface characteristics of the additively manufactured (AMed) metallic components [5]. The primary purpose of this post-processing technique is to remove any imperfections on the surface of the printed component, which may arise due to printing errors or poor

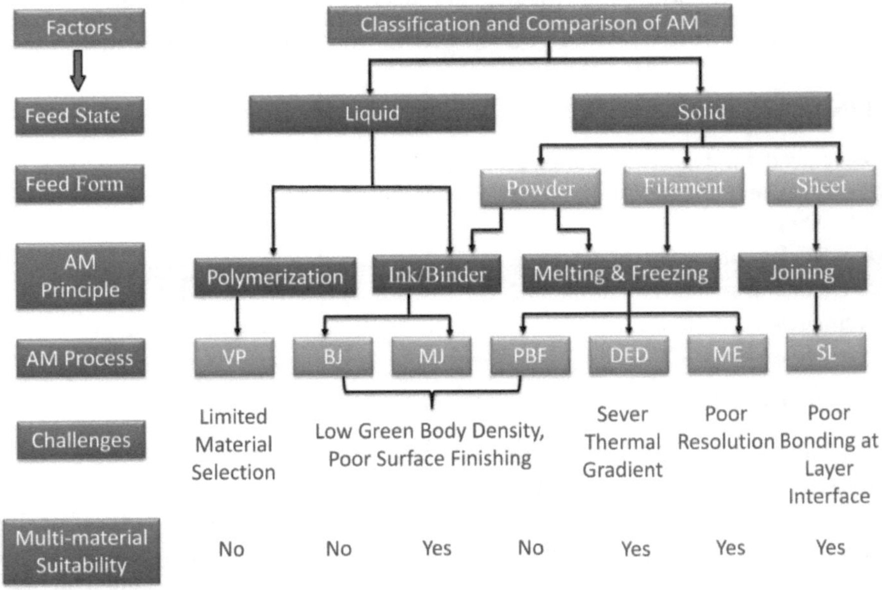

FIGURE 9.1 Classification of additive manufacturing processes (here, VP – vat photopolymerization, BJ – binder jetting, MJ – material jetting, PBF – powder bed fusion, DED – direct energy deposition, ME – material extrusion, SL – sheet lamination).

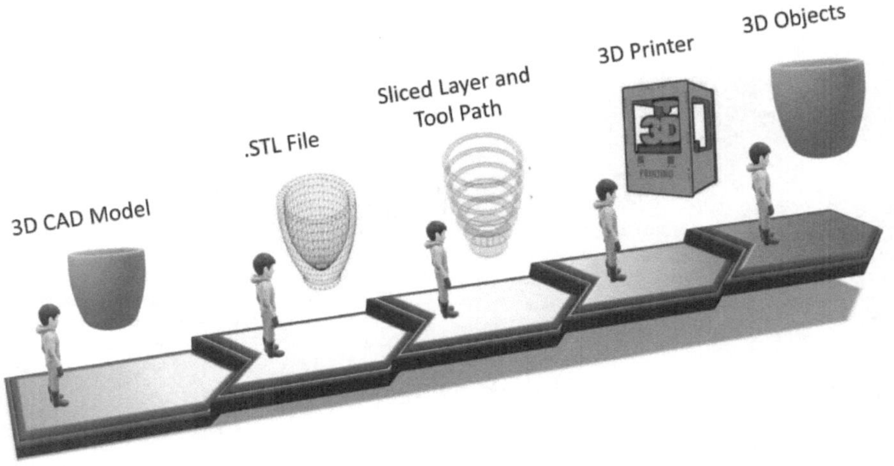

FIGURE 9.2 Various steps involved in the AM process.

material properties during printing, as well as to smooth and polish the surface of the components.

Based on the preceding discussion, the current study's purpose is to emphasize the problems associated with AM components and to demonstrate previous research work done to process metallic-based additive manufactured components using abrasive flow machining. The current chapter describes the requirements, classification of post-processing operations and the AFM process capabilities. Prior research on the AFM processing of metallic AMed components has also been explored. This will pave the way for researchers to further investigate the new hybrid machining methods for post-processing additive components at lower processing cost and time.

9.2 NEED AND CLASSIFICATION OF POST-PROCESSING PROCESSES

The main challenge with metal-based AM is the post-processing of components. Unlike traditional processed metals, the fabricated AMed metallic components consist of cracks, porosity, lack of green strength, anisotropic behavior, poor surface finishing, and stair-case defects [6,7]. The presence of these defects reduces the component's service life during application. Therefore, once the desired shape has been created, the component undergoes a post-processing step to ensure that it meets the required specifications. This step may involve smoothing the surface of the part, removing any excess material, and/or adding any necessary features. By carefully controlling the post-processing step, it is possible to create components with very tight tolerances and smooth surfaces. Nowadays various post-processing methods are utilized to enhance the surface characteristics of fabricated AM components. For

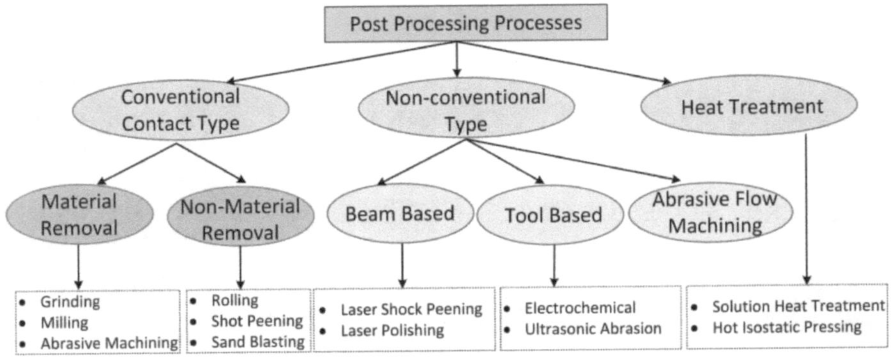

FIGURE 9.3 Classification of various AM post-processing processes.

the current work, post-processing techniques are classified as shown in Figure 9.3. The description of various post-processing techniques has been described in the next section.

9.3 ABRASIVE FLOW MACHINING

Abrasive flow machining (AFM) is a finishing process that uses abrasive laden slurry to smooth and polish surfaces. It is an effective alternative to traditional finishing processes like polishing, and grinding and can be used to produce finishes that are impossible to achieve with other methods. Recently, AFM has been used to finish a wide variety of AM materials, including metals, plastics, and composites. The schematic of the material removal mechanism and variant of AFM is shown in Figure 9.4. In this process, abrasive particles are suspended in a viscous carrier fluid, which flows under pressure through a narrow passage (Figure 9.4a). The abrasive particles in the fluid remove material from the surface as it flows past the work surface, leaving a smooth and polished finish. Depending on the media cylinder arrangements and movement direction of abrasive slurry, AFM is categorized into the four following variants:

(a) *One-way AFM:* In this type of AFM, the abrasive media pushed by hydraulic extrusion pressure flow only in one direction, as shown in Figure 9.4b.
(b) *Two-way AFM:* It consist of two oppositely arranged media cylinders that push the abrasive media in back and forth direction repeatedly, as shown in Figure 9.4c.
(c) *Multi-way AFM:* It consist of different media cylinders for feeding and receiving the abrasive media during the process, as illustrated in Figure 9.4d. The reduced processing time with better control over the process is the main advantage of this AFM variant.
(d) *Orbital AFM:* This type of AFM incorporates orbital motion in addition to axial movement of abrasive media into the system to enhance the interaction between abrasive media and workpiece, as depicted in Figure 9.4e.

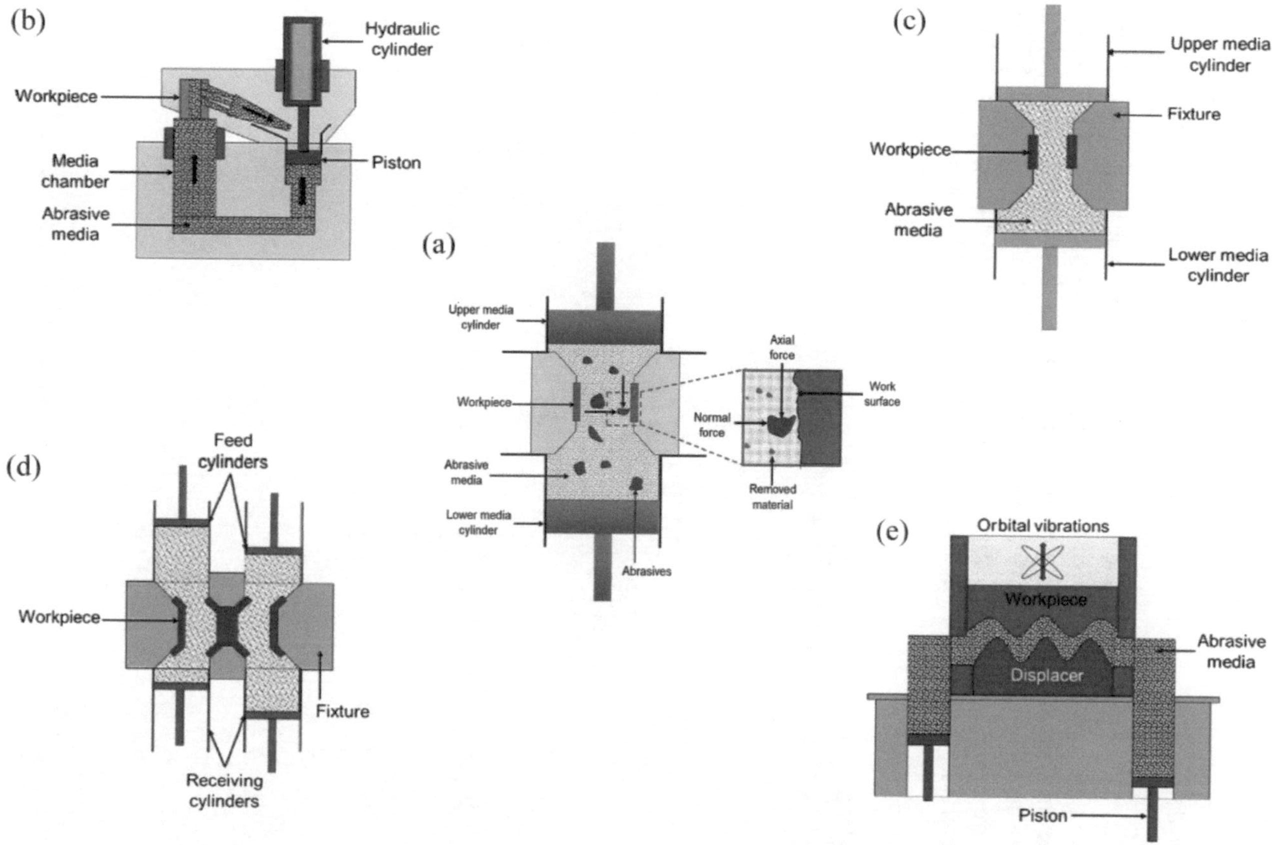

FIGURE 9.4 Schematic illustration of (a) AFM material removal mechanism, (b) one-way AFM, (c) two-way AFM, (d) multi-way AFM, and (e) orbital AFM [8].

9.4 REVIEW OF PAST WORK DONE

Additively manufactured metallic components generally have surface defects such as formation of oxide layer, inter-agglomerate porosity, anisotropy, cracking, balling, and high surface unevenness (5–15 μm) [9–12]. The common surface defects observed in the AMed components are shown in Figure 9.5. The presence of such defects makes the material prone to fatigue crack incubation and corrosion attacks. Galy et al. [13] carried out an in-depth investigation on the flaws of SLM parts and their repercussions and developed extensive tree diagrams to identify and represent the root cause of defects. Anisotropy, hot cracking, porosities, and poor surface finishing were identified as major root causes. For better understanding, various defects induced in AMed components are illustrated in Figure 9.5. The literature reveals various investigations were performed to improve the surface characteristics of AMed components by utilizing the AFM process. Peng et al.

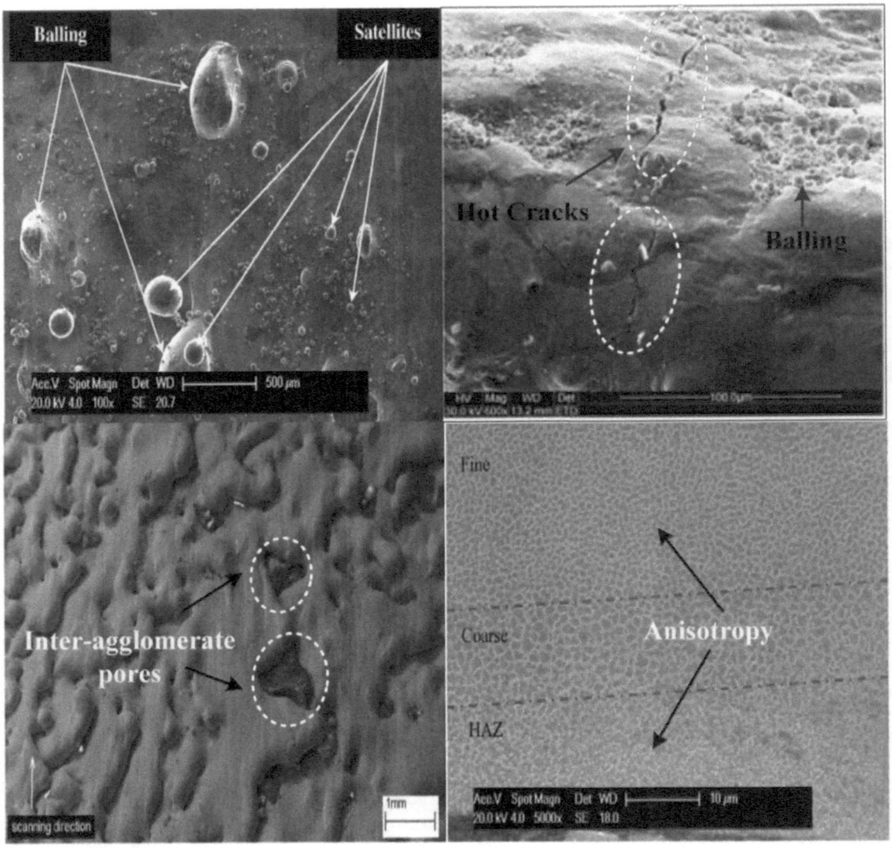

FIGURE 9.5 Various surface defects evident in AMed manufactured Al alloys [9–12].

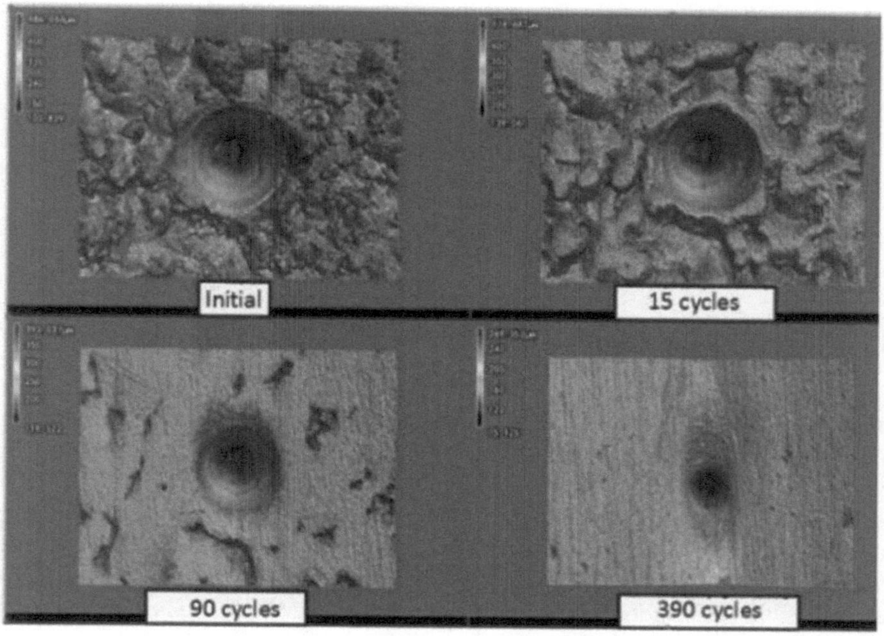

FIGURE 9.6 Surface topology of AMed components at different AFM cycles [14].

[14] utilized a two-way AFM method to post-process additively manufactured AlSi10Mg aluminum alloy. A micro-hole was drilled as a reference point to verify the surface integrity of the produced component, which was subsequently examined using a confocal laser scanning microscope. The surface topology of fabricated AMed components after 0, 15, 90, and 390 AFM cycles are represented in Figure 9.6. From this figure, it can be evident that only the small amount of surface adhered particles was removed during the first 15 cycles. The cluster of molten materials is removed after 90 cycles. However, the craters can be clearly observed, which were then totally diminished after 390 cycles. The component's surface roughness was observed to diminish as a result, going from an initial 14 μm Sa to 1.8 μm Sa.

In another work, Kum et al. [15] finished a nozzle guide vane (NGV) resembling laser sintered AM component. A medium viscosity media with SiC particles of #36 grit size was allowed to flow between the vanes with 500 mm/s average flow media velocity. The condition of AM component before and after the AFM process (as shown in Figure 9.7) reveals that the AFM can effectively improve the surface characteristics of the additively manufactured components.

Recently, the laser beam melting (LBM) process has emerged to fabricate waveguide channels for satellite communication applications. However, high transmission losses due to the rough internal surfaces of LBMed waveguides restrict the design engineers from using the AM technology for this application. To tackle this issue,

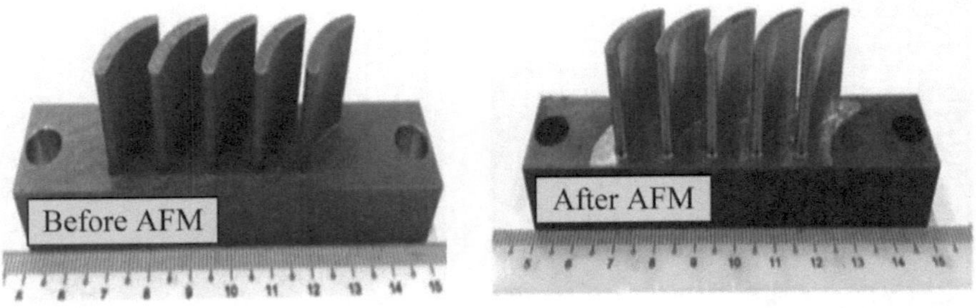

FIGURE 9.7 Surface condition of nozzle guide vanes before and after the AFM process [15].

François et al. [16] utilized AFM to improve the interior surface of Ti6Al4V and AlSi7Mg0.6 waveguides manufactured by LBM. Four waveguides were initially manufactured using the LBM process, which were subsequently finished by ULV 50–54%, extrude hone media. The surface topography of aluminum waveguide before and after the AFM is shown in Figure 9.8. The AFM finishing decreased the surface roughness of LBMed waveguides from 7–12 μm Ra to 0.7–3.9 μm Ra. Moreover, by calculating the attenuation coefficient (α), the waveguides' performance after AFM finishing was evaluated using the following equation (1):

$$\propto_a = 8668 \frac{\sqrt{\pi f \varepsilon}}{y\sqrt{K\sigma}} \cdot \frac{1+(2y/x)(f_c/f)^2}{\sqrt{1-(f_c/f)^2}} \tag{1}$$

where, f is frequency, ε is the material's dielectric permittivity, K is correction coefficient, y and x are cross-section dimension of fabricated waveguides, f_c is cut-off frequency, and σ is the material's electrical conductivity.

The AFM approach was also shown to be effective by Buchholz et al. [17] for enhancing the interior surface of a flow channel developed for an aerospike engine. Laser powder bed fusion (LPBF) was used to create the Inconel 718 flow channel, which was then inspected for surface quality. The mass flow rate of the fuel mixture (liquid oxygen + propellant ethanol) was found to be decreased due to the poor surface conditioning of the inside surface of the channel. The problem was fixed by applying the one-way AFM processing method on the manufactured channels, which decreased the Ra value from 92.3% to 95.1%. Similarly, in another study conducted by Mohammadian et al. [18] the interior surface of tubular IN625 component designed for the aerospace industry was improved by the innovative combined chemical-AFM process. The IN625 component was initially fabricated by laser powder bed-fused (LPBF) and subsequently finished by abrasive flow, chemical flow, and chemical-abrasive flow machining processes. The obtained results show that for part build orientations of 15° and 135°, respectively,

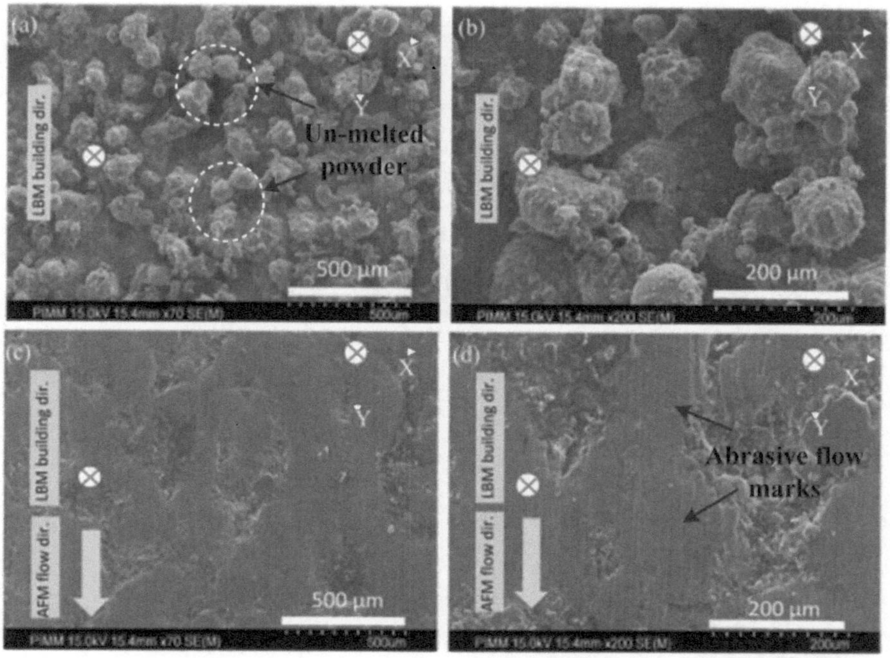

FIGURE 9.8 SEM image of internal surface of AlSi7Mg0.6 waveguide, (a, b) before AFM, and (c, d) after AFM [16].

chemical-abrasive flow machining reduced the surface roughness of the SLM-built part by 45% and 20%. Additionally, compared to chemical and abrasive flow procedures, the polishing time in chemical-abrasive flow machining was found to be 67% shorter.

Surface roughness and material removal rate are the backbone of the AFM process. However, most of the research work geared towards finishing the AMed components takes a model-specific approach. To combine the two models, i.e., surface roughness and material removal rate, Bouland et al. [19] conducted a research work in which laser powder bed-fused (LPBF) Ti6Al4V coupon containing multiple planer faces were finished by the AFM method. The surface characteristics of the component with different oriented surfaces before and after the AFM is depicted in Figure 9.9a–c. The figure represents that the AFM finishing effectively improved the surface topography of the LPBF component. Moreover, through experimental and simulation work, the combined graph of surface roughness and material removal was plotted, as shown in Figure 9.9d. Through this graph the machining allowance and number of cycles required for finishing the components with desired surface roughness outcomes can be determined. For instance, to get a surface (oriented at 135°) with a roughness of 5 μm, 25 finishing cycles are needed.

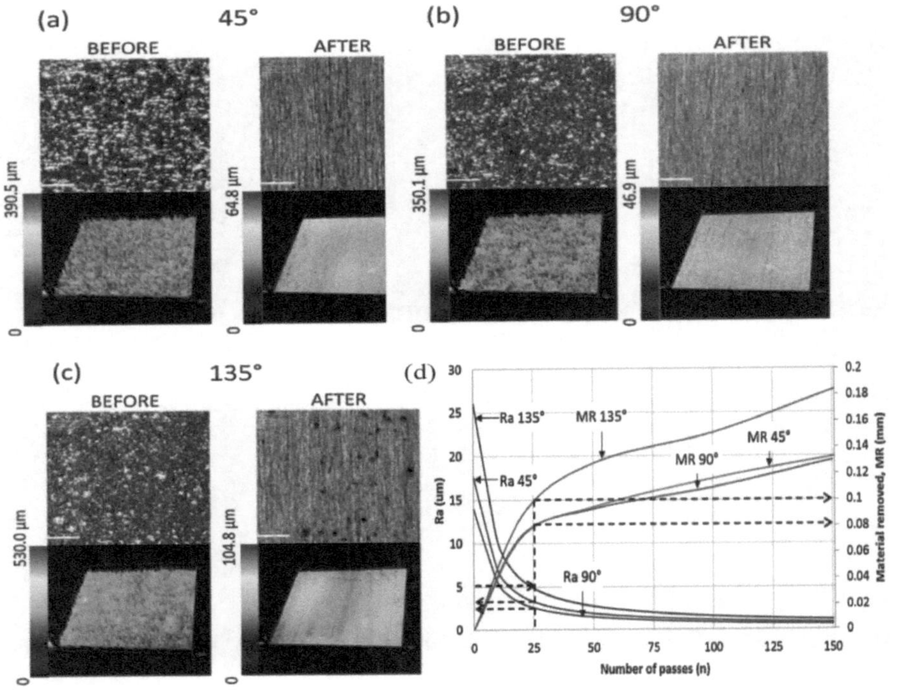

FIGURE 9.9 Surface characteristics of the LPBF component with three different oriented surfaces, (a) 45°, (b) 90°, and (c) 135°, and (d) material removal and surface roughness graph with respect to the number of passes [19].

9.5 CONCLUSIONS

The conclusions drawn from this study are as follows:

- Metal-based additive manufacturing processes can fabricate components with intricate geometries and desirable properties without requiring special tools. The fabricated components are utilized in many cutting-edge industries, such as aerospace, satellite communication, and bio-medical sectors.
- Additive manufacturing components consist of several surface defects, such as cracks, poor surface roughness, balling, and inter-agglomerate pores. These defects reduce the component's service life and applicability for intended applications.
- Abrasive flow machining (AFM) is a non-conventional finishing process that can improve the interior characteristics of AMed components. AFM is primarily used for polishing the cooling flow channels, mold's interior surfaces, and turbine blade surfaces.
- Chemical-abrasive flow machining is a novel hybrid finishing process that can improve the surface characteristics of AMed components with less time and better efficiency than chemical flow and abrasive flow processes.

REFERENCES

[1] Sharma S, Singh T, Dvivedi A Developments in tandem micro-machining processes to mitigate the machining issues at micron level: A systematic review, challenges and future opportunities. Mach Sci Technol 2022;26:515–70. https://doi.org/10.1080/10910344.2022.2129991

[2] Sharma S, Shamim FA, Dvivedi A, Kumar P, Singh T. Hybrid machining of metal matrix composites. Fabr. Mach. Adv. Mater. Compos., Boca Raton: CRC Press; 2022, p. 235–54. https://doi.org/10.1201/9781003327370-13.

[3] Mellor S, Hao L, Zhang D. Additive manufacturing: A framework for implementation. Int J Prod Econ 2014;149:194–201. https://doi.org/10.1016/j.ijpe.2013.07.008.

[4] Manfredi D, Calignano F, Ambrosio EP, Krishnan M, Canali R, Biamino S, et al. Direct Metal Laser Sintering: An additive manufacturing technology ready to produce lightweight structural parts for robotic applications. Metall Ital 2013;105:15–24.

[5] Duval-Chaneac MS, Han S, Claudin C, Salvatore F, Bajolet J, Rech J. Experimental study on finishing of internal laser melting (SLM) surface with abrasive flow machining (AFM). Precis Eng 2018;54:1–6. https://doi.org/10.1016/j.precisioneng.2018.03.006.

[6] Shiyas KA, Ramanujam R. A review on post processing techniques of additively manufactured metal parts for improving the material properties. Mater Today Proc 2021;46:1429–36. https://doi.org/10.1016/j.matpr.2021.03.016.

[7] Mahmood MA, Chioibasu D, Ur Rehman A, Mihai S, Popescu AC. Post-processing techniques to enhance the quality of metallic parts produced by additive manufacturing. Metals (Basel) 2022;12:77. https://doi.org/10.3390/met12010077.

[8] Dixit N, Sharma V, Kumar P. Research trends in abrasive flow machining: A systematic review. J Manuf Process 2021;64:1434–61. https://doi.org/10.1016/j.jmapro.2021.03.009.

[9] Aboulkhair NT, Maskery I, Tuck C, Ashcroft I, Everitt NM. On the formation of AlSi10Mg single tracks and layers in selective laser melting: Microstructure and nano-mechanical properties. J Mater Process Technol 2016;230:88–98. https://doi.org/10.1016/j.jmatprotec.2015.11.016.

[10] Gu D, Shen Y. Balling phenomena in direct laser sintering of stainless steel powder: Metallurgical mechanisms and control methods. Mater Des 2009;30:2903–10. https://doi.org/10.1016/j.matdes.2009.01.013.

[11] Simchi A. Direct laser sintering of metal powders: Mechanism, kinetics and microstructural features. Mater Sci Eng A 2006;428:148–58. https://doi.org/10.1016/j.msea.2006.04.117.

[12] Aboulkhair NT, Everitt NM, Ashcroft I, Tuck C. Reducing porosity in AlSi10Mg parts processed by selective laser melting. Addit Manuf 2014;1:77–86. https://doi.org/10.1016/j.addma.2014.08.001.

[13] Galy C, Le Guen E, Lacoste E, Arvieu C. Main defects observed in aluminum alloy parts produced by SLM: From causes to consequences. Addit Manuf 2018;22:165–75. https://doi.org/10.1016/j.addma.2018.05.005.

[14] Peng C, Fu Y, Wei H, Li S, Wang X, Gao H. Study on improvement of surface roughness and induced residual stress for additively manufactured metal parts by abrasive flow machining. Procedia CIRP 2018;71:386–9. https://doi.org/10.1016/j.procir.2018.05.046.

[15] Kum CW, Wu CH, Wan S, Kang CW. Prediction and compensation of material removal for abrasive flow machining of additively manufactured metal components. J Mater Process Technol 2020;282. https://doi.org/10.1016/j.jmatprotec.2020.116704.

[16] François M, Han S, Segonds F, Dupuy C, Rivette M, Turpault S, et al. Electromagnetic performance of Ti6Al4V and AlSi7Mg0.6 waveguides with laser beam melting (LBM) produced and abrasive flow machining (AFM) finished internal surfaces. J Electromagn Waves Appl 2021;35:2510–26. https://doi.org/10.1080/09205071.2021.1954554.

[17] Buchholz M, Gruber S, Selbmann A, Marquardt A, Meier L, Müller M, et al. Flow rate improvements in additively manufactured flow channels suitable for rocket engine application. CEAS Sp J 2022. https://doi.org/10.1007/s12567-022-00476-7.

[18] Mohammadian N, Turenne S, Brailovski V. Surface finish control of additively-manufactured Inconel 625 components using combined chemical-abrasive flow polishing. J Mater Process Technol 2018;252:728–38. https://doi.org/10.1016/j.jmatprotec.2017.10.020.

[19] Bouland C, Urlea V, Beaubier K, Samoilenko M, Brailovski V. Abrasive flow machining of laser powder bed-fused parts: Numerical modeling and experimental validation. J Mater Process Technol 2019;273. https://doi.org/10.1016/j.jmatprotec.2019.116262.

10 An Overview of the Quality Characteristics Challenges in Additive Manufacturing

B. Özbay Kısasöz[1] *and A. Kısasöz*[2]
[1]Aluminum Test Training and Research Center (ALUTEAM), Fatih Sultan Mehmet Vakif University, Istanbul, Turkey
[2]Department of Metallurgical and Materials Engineering, Yildiz Technical University, Istanbul, Turkey

CONTENTS

10.1 INTRODUCTION

Recently, additive manufacturing (AM), commonly known as three-dimensional (3D) printing, has become an environmentally friendly technology that provides energy savings, less material consumption and efficient production. Additive manufacturing is a small amount of material at a time machining is a process that involves joining successive parts directly from a computer-aided-design (CAD) model to produce the desired part (Ngoa et al., 2018).

3D additive manufacturing technologies are generally classified according to the technology used, materials, and process types (Figure 10.1). According to the ISO/ASTM 529000:2015 standard, 3D additive manufacturing methods are classified into seven categories. Although each method has advantages and disadvantages in different aspects, it appeals to different areas of use. The products produced by the additive manufacturing method are mainly used in the aerospace, aviation, defense, medicine, electrical-electronics, and automotive industries. (Hajare and Gajbhiye, 2022).

DOI: 10.1201/9781003276111-10

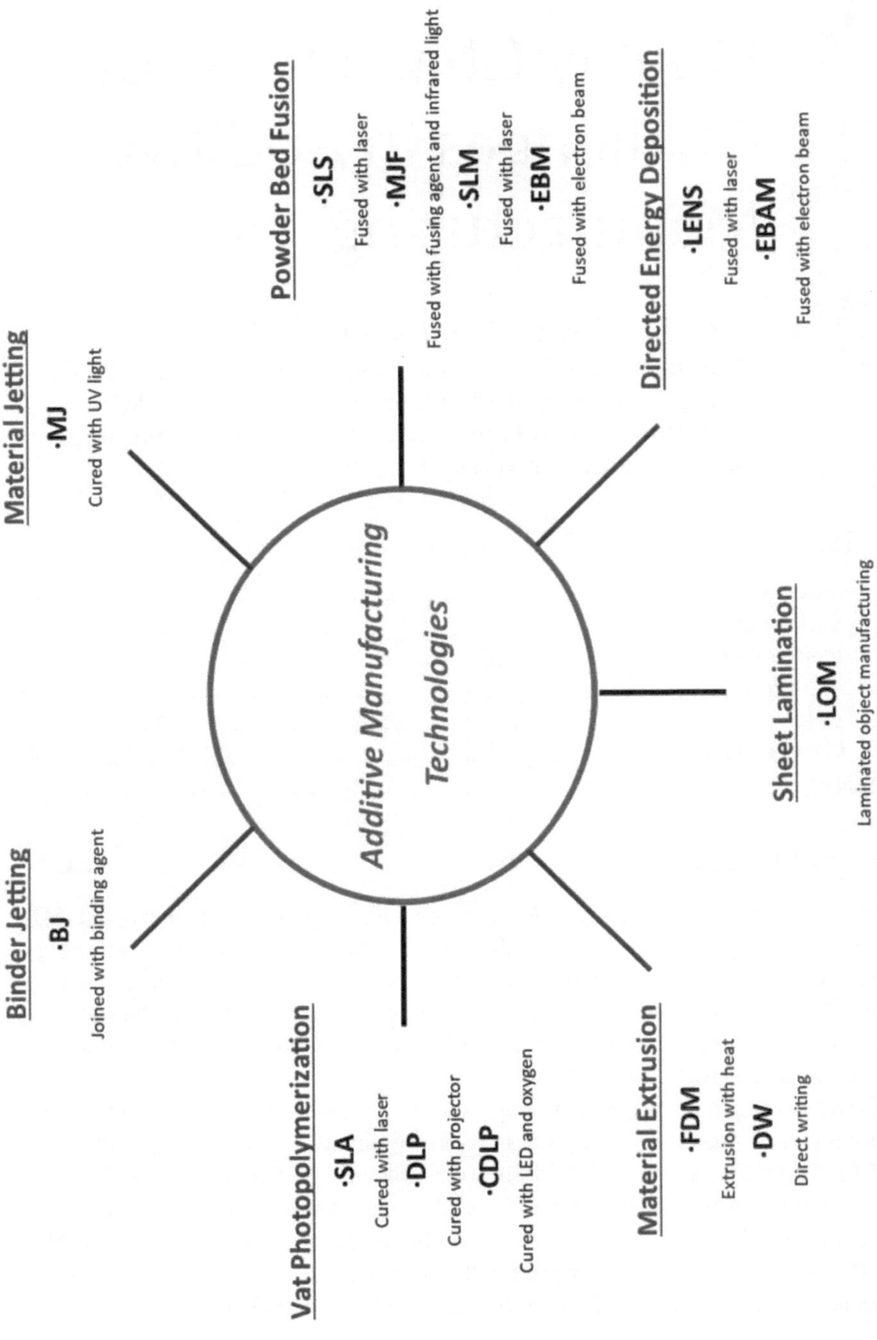

FIGURE 10.1 Seven categories of additive manufacturing (Rafiee et al., 2020).

The growing interest in studies about the AM techniques provide the adaptation of AM to many industrial products in various dimensions, shapes, and designs. Compared to conventional manufacturing methods, the AM techniques ensure several advantages like flexibility in design, obtaining complex geometry with high precision and material savings. AM techniques can be used for a wide range of materials, including polymers, ceramics, and metals. Moreover, composite materials can also be fabricated by using various reinforcements like particulates and fibers. On the other hand, various problems and challenges may arise in the usage of AM methods. Material properties and the quality of the final products are affected negatively by these problems and challenges. This chapter discusses the challenges in AM in terms of surface quality, staircase and edge effects, molten pool, and defect formation aspects.

10.2 SURFACE QUALITY

The additive manufacturing method operates in many fields ranging from aviation to biomedicine. The manufactured parts must have a surface quality that can meet the demands and requests of these different sectors.

The surface quality and surface roughness approach display differences in additive manufacturing according to various raw materials and production methods. Surface texture problems are typical drawbacks of laser powder bed fusion systems. Therefore, powdered-form materials induce relatively rough structure surfaces while layer-by-layer LS processing (Kumbhar and Mulay, 2015; Metelkova et al., 2021).

The original powder geometry may be noticeable to some extent on the surface, and staircase effects can result in inhomogeneous, stepped surfaces that are visible due to (incorrect) part orientations. Nevertheless, by optimizing the process parameters of the SLS method, the quality of end products can be enhanced. Various studies are available to obtain better surface quality in pre-production (production planning) and post-production steps with optimization of the process parameters (Schmid, 2018; Hashmi et al., 2021). Moreover, several investigations have been carried out related to surface quality and process parameters' correlation. Negi et al. (Negi et al. 2014) examined the effects of process temperature, laser power, laser scan speed, and spacing on surface roughness process parameters for SLS-fabricated glass-filled polyamide parts. It has been stated that the most important factor, following the laser power, is the scanning range, which contributes the most to improving the surface roughness. The optimum values of different parameters have been obtained and verified by conducting verification experiments. There was also an investigation and comparison by ANOVA, and according to the obtained results, the most important values for surface properties are laser power and scan speed, followed by the scan spacing. Furthermore, experimental test results concluded that the surface roughness values were reduced by increasing laser power and scan length values while enhanced by increasing scan speed and spacing.

A research revealed the surface quality of printed porous materials, which were produced by various AM methods (SLA, MJF, and FDM 3DP). The outcomes show that SLA demonstrated the best surface finish at the minimum standard deviations of the roughness values. Besides, the mean roughness (Sa) and root mean square (RSM) roughness (Sq) values of better surface finish of MJF samples are lower than the same values of the FDM method (Bodaghi et al., 2022). In another study, a laser-assisted finishing process was done to develop the surface quality of FDM parts. The comparison of the results demonstrated that when the laser-based finishing process is performed, the most proper surface finishing circumstances are as follows: (i) low arithmetic surface roughness (Ra), (ii) negative skewness (Rsk), (iii) kurtosis value (Rku)>3 (Taufik et al., 2017).

On the other hand, although it is not possible to obtain a surface smoothness as in CNC machines in terms of metal materials, the surface roughness challenge should be minimized by considering different aspects of products produced with additive manufacturing. Surface roughness is also a very important factor like surface form, waviness, and hardness; it is highly dependent on various parameters (powder particle geometries and sizes, layer thickness, wall angle, melt pool size, etc.). The laser remelting (LSR) technique can be an easy solution without removing the part from the build platform and avoiding fixing errors. For this purpose, it can be stated that the LSR process provides a unit size refinement and homogeneity of the microstructure on the surface (Simoni et al. 2021).

10.3 STAIRCASE AND EDGE EFFECTS

Due to the fact that AM is a layer-by-layer process, a staircase effect may appear on the slanted surfaces of the final parts. This effect (Figure 10.2), which is one of the most important factors affecting the character of additive manufacturing parts, influences surface roughness at micro and macro levels.

The staircase is responsible for the surface roughness of the AM parts, and the increasing staircase effect results in higher surface roughness (Simoni et al., 2021). Moreover, achieving the desired surface quality requires optimizing the staircase formation. The staircase effect can depend on the following parameters: layer thickness,

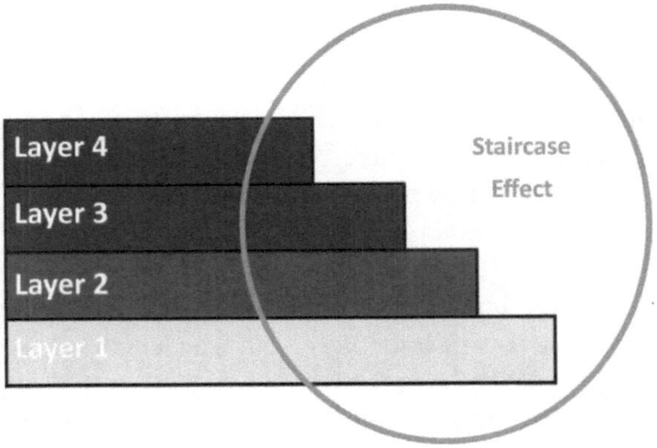

FIGURE 10.2 Schematic representation of the "staircase effect".

layer shrinkage, laser angle, and build orientation. The layer thickness is the most critical parameter of the staircase characteristics. Changes in the layer thickness also cause a significant effect on the occurrence of the staircase effect, and the decrease in the layer thickness minimizes the formation of stair-stepping (Charles et al., 2019). Moreover, increasing the wall angle reduces the staircase effect. Build orientation also has an influence on the staircase formation and accordingly affects the surface roughness of the final parts (Das et al., 2015). Moreover, shrinkage of the layers, deposition of the layers from the bottom up and usage of incorrect laser angle arise the staircase effect (Matos et al., 2020). Besides, the ripple effect triggers the staircase effect.

Enhanced scan speed causes a decrease in heat input, and it is insufficient to complete the wetting of the surrounding powder particulates. Insufficient wetting results in the formation of balling, and the balling leads to the formation of a ripple effect. Staircase structure is observed in the AM parts with the formation of the ripple effect (Shen et al., 2006). It was proved that higher scan speed values induce a balling in the SLM process of the 316L (Yadroitsev et al., 2013). During the AM processing of the parts, the feedstock is melted, and each melted layer solidifies. Barely, partially melted feedstock sticks to the edge of the melt track, which induces an edge effect that causes an increase in surface roughness. Also, pores can form between the partially melted powders and the molten metal, and spherical pores can be observed along the edges of the melt track. These pores cause a stress concentration, act as a preferential site for crack initiation, and reduce the mechanical properties of the final parts (Sola and Nouri, 2019).

10.4 MELT POOL

The additive manufacturing (AM) techniques use high energy density values to obtain suitable final products. An electric arc, laser beam, and electron beam can be used as energy sources (Sames et al., 2016). In these methods, powders are melted by the transferred energy from the source to the deposited layer, and the melt is deposited on the produced layer. Each melting step results in the formation of a melt pool. Parts are produced by repeated melting and solidification-based processes (Figure 10.3). As a result, the whole techniques include melting and solidification processes during the fabrication of engineering parts.

Multilayered components that can be produced by the repeated traverse of the laser or electron beams. The power source movement provides the formation of melt pools, and the solidification of the melt pools enables the production of the components. Generally, small melt pool size and high peak temperature are the main characteristics of the laser beam and electron beam-based processes, and thus, rapid solidification occurs due to these characteristics (Manvatkar et al., 2015). Rapid solidification requires the control of many variables like scanning speed, source power, power density, hatch spacing, chemical composition, particulate size distribution and flow rate of the powders in order to produce high-quality components. The microstructure of the AM parts can be controlled by the mentioned parameters, and desired mechanical properties are achieved (Mondal et al., 2020).

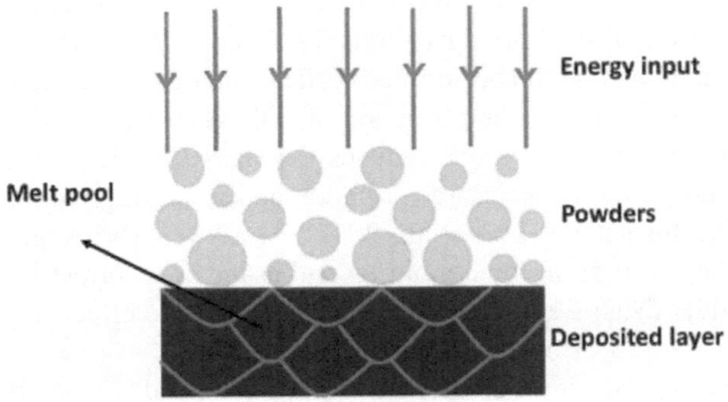

FIGURE 10.3 Schematic illustration of the melt pool formation.

Besides the rapid solidification characteristic, the cyclic nature of the AM techniques causes continuous changes in the solidification behavior of the components during the AM process. The thermal gradient (*G*) and the solidification velocity (*R*) vary continuously, and the melt pool properties differ between the initial and final layers. The columnar grain structure occurs at the beginning of the scan due to epitaxial growth along the building direction (Basak and Das, 2016). The grains also transform into equiaxed due to the progressive heating effect at the end of the scan. The columnar grains deteriorate the mechanical properties of the AM parts. Thus, one of the basic criteria in designing the AM parameters is obtaining an equiaxed melt pool structure to provide higher mechanical properties. The microstructure of the products can be controlled by changing the laser power, scan speed, and substrate temperature (Rao et al., 2016). Moreover, Hadadzadeh et al. revealed that building direction is critical for microstructural properties. The grain structure changes from columnar to equiaxed by changing the building direction from vertical to horizontal (Hadadzadeh et al., 2018).

The shielding gas also significantly influences the development of the molten pool quality. Electron-beam AM method is generally carried out in a vacuum environment. Thus, the shielding gas effect can be ignored, unlike the other methods. In general, an inert gas such as He is used in the AM processes. It is known that inert gasses are non-reactive and insoluble in molten metal. Therefore, unless the shielding gas could not escape out of the molten pool, the shielding gas will be trapped in the structure, and the pores will occur in the melt pool. Non-inert shielding gasses like nitrogen may be preferred to prevent pore formation. The nitrogen dissolves in the liquid metal before the solidification, and the porosity will be reduced or eliminated. However, since the gases are reactive, they cannot be used in the additive manufacturing process of all metals, and their usage is limited. Accordingly, AM of metals that are prone to porosity formation should be carried out in a vacuum environment as much as possible (Thijs et al., 2010).

The melt pool characteristics are critical for optimizing the process conditions because melt pool properties directly affect the quality of AM parts. Therefore, the melt pool properties have to be considered carefully. Thus, studies have been carried out on the numerical modeling of the melt pool characteristic in recent years.

Numerical modeling is a powerful tool to help understand the solidification of the melt pool, heat generation, heat and fluid flow in AM process. The numerical models generally include predicting melt pool geometry and temperature distribution in the deposited layer. Moreover, the models must focus on defects like porosity, lack of fusion, and spattering, etc. (Schoinochoritis et al., 2017). In addition, the numerical models are based on fluid flow and heat transfer and their physical theories. The model that presents a correct approach in terms of the mentioned parameters (Cook and Murphy, 2020): heating and melting of the initial powders by arc energy, laser or electron beam, surface tension and wetting ability between the substrate and partially melted powders, formation of the melt pool and liquid metal and the occurrence of the porosity, evaporation of liquid owing to high energy input characteristics of the process, heat and fluid flow in the melt pool, radiative and convective heat transfer from the metal surfaces, phase transformations of metal between liquid and solid with associated latent heat release or absorption.

It is clear that determining the melt pool characteristics and revealing the influence of process parameters on melt pool properties are considerable for AM parts. Numerical modeling has great potential in that manner. Researches in the modeling of the AM process are increasing day by day, and the models have been developed to solve problems such as necessary to complicated physics for melt pool analysis and extensive computational resources for advanced models.

10.5 DEFECTS

The formation of defects is the major concern for the AM processes. As mentioned before, various production methods can be used to fabricate materials through the additive manufacturing process. Material type and the AM method can influence the defect type and concentration.

One of the main problems encountered in the additive manufacturing of polymers is the defects that occur due to the presence of oxygen in the environment and the moisture absorption tendency of the polymer. In additive manufacturing of polymers such as ABS and PA, significant increases in mechanical values are obtained by removing the oxygen from the process (Lederle et al., 2016). In AM methods using filaments like FDM, the moisture content increases as the length of the filament fed into the system increases. Correspondingly, the mechanical properties of polymer-based AM parts decrease by moisture absorption. Especially the thermal effect in AM causes expansion of the absorbed moisture, which results in porosity formation (Goh et al., 2020). Polymers exhibit lower mechanical properties in engineering applications. Thus, polymer-based composites have been developed to meet mechanical expectations. The usage of AM techniques in polymer-based composite production is increasing, and various challenges/defects can be encountered during the process. Particulates, short and continuous fibers are used as reinforcements for improving the mechanical behavior of the polymers produced by AM. Pull-outs, de-bonding, and breakage of the fibers are the main challenges for the fiber-reinforced AM products. Fiber orientation is also a critical factor for the mechanical characteristics of the final product. Improved mechanical properties like tensile strength and toughness can be obtained in 0° orientation, whereas 90° orientation causes the occurrence of the weakest mechanical properties (Goh et al., 2019). Researchers have reported void formation due to the usage of fiber reinforcement besides the fiber orientation problem (Saroia et al., 2020). Particulate-reinforced polymers have been developed in order to overcome the challenges of fiber-reinforced polymer composites in AM techniques. Various reinforcements like copper, hollow glass microspheres, and silicon carbide are used as particulate reinforcements (Özbay Kısasöz et al., 2022). These lower the anisotropy, improve the mechanical behavior and flowability and possess flame-retardancy and electrical conductivity to the polymer structure. For instance, the use of expandable microspheres leads to a decrease in porosity fraction, and the mechanical properties of the AM part will be improved (Wang et al., 2016). However, achieving a homogenous distribution of the particulates is the main challenge in manufacturing particulate-reinforced polymer matrix composites by AM techniques. One of the solutions to overcome this challenge is the usage of coated reinforcements (Wencke et al., 2021).

Ceramic materials possess higher mechanical strength and hardness compared to metals and polymers (Kisasoz et al., 2011). Nevertheless, the brittleness of ceramic-based materials is the main challenge for engineering applications. Various problems are encountered in the production of ceramics by traditional production methods because the process consists of many production steps and the brittleness of the material (Travitzky et al., 2014). AM techniques can be solutions to the difficulties encountered in traditional manufacturing. However, defects such as cracks and porosity may be encountered in the AM of ceramic-based materials. For instance, ceramic-based AM parts can include a higher porosity fraction by producing with the SLS technique. Hence, the SLS of ceramic materials requires a post-sintering process.

In contrast, porous AM ceramic parts can be advantageous for applications requiring specific strength. In this case, porous ceramics can be produced with AM techniques by porosity-forming additives (Li et al., 2016). Moreover, pre-heating has to be applied for AM of ceramic parts in order to prevent shrinkage and crack formations. It can be seen in the literature that ZrO_2 and Y_2O_3 mixtures were pre-heated at 1600 °C (Wilkes et al., 2013) and 1715 °C (Shishkovsky et al., 2007) in the AM process. Due to the high melting temperatures of the ceramics, melting problems may occur during the AM process of the ceramic-based parts. Therefore, binders are added into feedstock in order to reduce the melting temperature of the powder mixtures (Derby, 2015).

Metallic materials can be fabricated by various AM techniques such as powder bed systems, powder feed systems, and wire feed systems (Frazier, 2014). In the production of metals with the AM techniques, various defects such as porosity, cracks, and impurities may occur depending on the alloy and techniques. Porosity is one of the most crucial problems that can be encountered in the AM of metals. The mechanical behavior of the metals is significant in engineering applications, and porosity formation reduces the mechanical properties of metallic AM parts. The porosity reduces the load-carrying capacity of the structure, and it acts as a preferential site for crack initiation. The porosity structure can be divided into two groups in AM of metallic parts: (i) gas porosity and (ii) lack of fusion (Honarvar and Varvani-Farahanib,

2020). Gas pores can form via a chemical reaction in the molten metal, supersaturation of the dissolved gas and gas entrapment during the process. Generally, gas pores are characterized by their spherical shape. Higher solidification rates prevent the removal of gas through the molten metal, and the formation of entrapped gas takes place. However, lower solidification rates allow the growth and coalescence of the pores. Thus, cooling and solidification rates are the critical parameters for the AM of metallic parts. The lack of fusion is a larger defect compared to gas porosity, and also, their size can differ from 50 μm to several millimeters. Insufficient overlap between passes and increasing hatch spacing result in a lack of fusion defects (Brennan et al., 2021). It was proved that increasing hatch spacing from 50 to 100 μm promotes the formation of a lack of fusion defect in the SLM process of the Ti6Al4V (Thijs et al., 2010). Solidification is one of the main parameters for the AM of metal parts. The AM process parameters and solidification characteristics of the alloy have a significant influence on the formation of solidification cracks. A lower scan speed and applied preheating lead the slower cooling rates and reduced crack formation tendency (Buchbinder et al., 2015). Moreover, the alloy with a wider solidification temperature range shows more susceptibility to forming solidification cracks. The wide solidification temperature range increases the accumulation of thermal strain during the solidification, resulting in the crack formation tendency (Lippold, 2015). Impurities can form due to the excessive content of the insoluble elements in the AM environment. In particular, the existence of oxygen in the production environment causes the oxidation, formation of impurities, and contamination of the feedstock. Impurities lead to an increase in the pore fraction of the AM parts, and the ductility of the structure is reduced (Shiva et al., 2015). Hence, the usage of an inert environment or shielding gas like argon is critical for the AM of metal-based parts in terms of obtaining defect-free structures and providing desired microstructural and mechanical properties.

In AM techniques, desired properties can be obtained by the production parts with reduced defect fraction. Various defects can form in the structure, and the precautions may differ due to AM techniques and feedstock properties. For this reason, appropriate process parameters should be defined to prevent the formation of possible defects considering the technique, feedstock, and process conditions.

10.6 CONCLUSIONS

The AM methods have great potential for engineering applications owing to providing high-quality final products. However, the AM techniques cause various challenges depending on the method, process parameters, and material. These challenges have a negative effect on the quality of the AM parts. In this study, the most common challenges of the AM have been discussed, and suggestions are presented. The interest in AM techniques has been growing recently, and the studies have focused on the challenges and their solutions. In addition, using the appropriate technique, parameters, and feedstock is still the most important factor for achieving the greatest final part quality.

REFERENCES

Basak, A., and Das, S. (2016). Epitaxy and microstructure evolution in metal additive manufacturing. *Annual Review of Materials Research*, 46, 125–149.

Bodaghi, M., Mobin, M., Ban, D., Lomov, S. V., and Nikzad, M. (2022). Surface quality of printed porous materials for permeability rig calibration. *Materials and Manufacturing Processes*, 37(5), 548–558.

Brennan, M. C., Keist, J. S., and Palmer, T. A. (2021). Defects in metal additive manufacturing processes. *Journal of Materials Engineering and Performance*, 30, 4808–4818.

Buchbinder, D., Meiners, W., and Wissenbach K. (2015). Selective laser melting of aluminum die-cast alloy—Correlations between process parameters, solidification conditions, and resulting mechanical properties. *Journal of Laser Applications*, 2015, 27, S29205.

Charles, A., Elkaseer, A., Thijs, L., Hagenmeyer, V., and Scholz, S. (2019). Effect of process parameters on the generated surface roughness of down-facing surfaces in selective laser melting. *Applied Sciences*, 9(6), 1256.

Cook, P. S., and Murphy, A. B. (2020). Simulation of melt pool behaviour during additive manufacturing: Underlying physics and progress. *Additive Manufacturing*, 31, 100909.

Das, P., Chandran, R., Samant, R., and Anand, S. (2015). Optimum part build orientation in additive manufacturing for minimizing part errors and support structures. *Procedia Manufacturing*, 1, 343–354.

Derby, B. (2015). Additive manufacture of ceramics components by inkjet printing. *Engineering*, 1(1), 113–123.

Frazier, W. E. (2014). Metal additive manufacturing: A review. *Journal of Materials Engineering and Performance*, 23, 1917–1928.

Goh, G. D., Yap, Y. L., Agarwala, S., and Yeong, W. Y. (2019). Recent progress in additive manufacturing of fiber reinforced polymer composite. *Additive Materials Technologies*, 4(1), 1800271.

Goh, G. D., Yap, Y. L., Tan, H. K. J., Sing, S. L., Goh, G. L., and Yeong, W. Y. (2020). Process–structure–properties in polymer additive manufacturing via material extrusion: A review. *Critical Reviews in Solid State and Materials Sciences*, 45, 113–133.

Hadadzadeh, A., Amirkhiz, B. S., Li, J., and Mohammadi, M. (2018). Columnar to equiaxed transition during direct metal laser sintering of AlSi10Mg alloy: effect of building direction. *Additive Manufacturing*, 23, 121–131.

Hajare, D. M., and Gajbhiye, T. S. (2022). Additive manufacturing (3D printing): Recent progress on advancement of materials and challenges. *Materials Today: Proceedings*, 58, 736–743.

Hashmi, A. W., Mali, H. S., and Meena A. (2021). Improving the surface characteristics of additively manufactured parts: A review. *Materials Today: Proceedings*, https://doi.org/10.1016/j.matpr.2021.04.223.

Honarvara, F., Varvani-Farahanib, A. (2020). A review of ultrasonic testing applications in additive manufacturing: Defect evaluation, material characterization, and process control. *Ultrasonics*, 105, 106227.

Kisasoz, A., Guler, K. A., and Karaaslan, A. (2011). Fabrication and characterization of SiC preforms for metal matrix composites. *Materials Testing*, 53(1), 634–637.

Kumbhar, N. N., and Mulay, A. V. (2015). Post processing methods used to improve surface finish of products which are manufactured by additive manufacturing technologies: A review. *Journal of the Institution of Engineers (India): Series C*, 99(4), 481–487.

Lederle, F., Meyer, F., Brunotte, G. P., Kaldun, C., and Hübner, E. G. (2016). Improved mechanical properties of 3D printed parts by fused deposition modelling processed under the exclusion of oxygen. *Progress in Additive Manufacturing*, 1, 3–7.

Li, X., Gao, M, and Jiang, Y. (2016). Microstructure and mechanical properties of porous alumina ceramic prepared by a combination of 3–D printing and sintering. *Ceramics International*, 42(10), 12531–12535.

Lippold, J. C. (2015). *Welding Metallurgy and Weldability*. Hoboken: Wiley.

Manvatkar, V., De, and DebRoy, T. (2015). Spatial variation of melt pool geometry, peak temperature and solidification parameters during laser assisted additive manufacturing process, *Materials Science and Technology*, 31, 8, 924–930.

Matos, M. A., Rocha, A. M. A. C., and Pereira, A. I. (2020). Improving additive manufacturing performance by build orientation optimization. *The International Journal of Advanced Manufacturing Technology*, 107, 1993–2005.

Metelkova, J., Vanmunster, L., Haitjema, H., and Hooreweder, B. V. (2021). Texture of inclined up-facing surfaces in laser powder bed fusion of metals. *Additive Manufacturing*, 42, 101970.

Mondal, S., Gwynn, D., Ray, A., and Basak, A. (2020). Investigation of melt pool geometry control in additive manufacturing using hybrid modelling. *Metals*, 10, 683.

Negi, S., Dhiman, S., and Sharma, R. K. (2014). Investigating the surface roughness of SLS fabricated glass-filled polyamide parts using response surface methodology. *Arabian Journal of Science and Engineering*, 39, 9161–9179.

Ngoa, T. D., Kashania, A., Imbalzanoa, G., Nguyena, K. T. Q., and Hui, D. (2018). Additive manufacturing (3D printing): A review of materials, methods, applications and challenges. *Composites Part B*, 143, 172–196.

Özbay Kısasöz, B., Serhatlı, I. E., Bulduk, M. E. (2022). Selective laser sintering manufacturing and characterization of lightweight PA 12 polymer composites with different hollow microsphere additives. *Journal of Materials Engineering and Performance*, 31(5), 4049–4059.

Rafiee, M., Farahani, R. D., and Therriault, D. (2020). Multi-material 3D and 4D printing: A survey. *Advanced Science*, 7(12), 1902307.

Rao, H., Giet, S., Yang, K., Wu, X., and Davies, C. H. J. (2016). The influence of processing parameters on aluminium alloy A357 manufactured by selective laser melting. *Materials and Design*, 109, 334–346.

Sames, W. J., List, F. A., Pannala, S., Dehoff, R. R., and Babu, S. S. (2016). The metallurgy and processing science of metal additive manufacturing. *International Materials Reviews*, 61, 315–360.

Saroia, J., Wang, Y., Wei, Q., Lei, M., Li, X., Guo, Y., and Zhang, K. (2020). A review on 3D printed matrix polymer composites: Its potential and future challenges. *The International Journal of Advanced Manufacturing Technology*, 106, 1695–1721.

Schmid, M. (2018). *Laser Sintering with Plastics: Technology, Processes, and Materials*. Munich, Germany: Hanser.

Schoinochoritis, B., Chantzis, D., and Salonitis, K. (2017). Simulation of metallic powder bed additive manufacturing processes with the finite element method: A critical review. *Journal of Engineering Manufacture*, 23, 1, 96–117.

Shen, Y. F., Gu, D. D., and Pan, Y. F. (2006). Balling process in selective laser sintering 316 stainless steel powder. *Key Engineering Materials*, 315, 357–360.

Shishkovsky, I., Yadroitsev, I., Bertrand, P., and Smurov, I. (2007). Alumina–zirconium ceramics synthesis by selective laser sintering/melting. *Applied Surface Science*, 254(4), 966–970.

Shiva, S., Palani, I. A., Mishra, S. K., Paul, C. P. and Kukreja, L. M. (2015). Investigations on the influence of composition in the development of Ni-Ti shape memory alloy using laser based additive manufacturing. *Optics and Laser Technology*, 69, 44–51.

Simoni, F., Huxol, A., and Villmer, F. J. (2021). Improving surface quality in selective laser melting based tool making. *Journal of Intelligent Manufacturing*, 32, 1927–1938.

Sola, A., and Nouri, A. (2019). Microstructural porosity in additive manufacturing: The formation and detection of pores in metal parts fabricated by powder bed fusion, *Journal of Advanced Manufacturing and Processing*, 2019, 1, 10021.

Taufik, M., and Prashant, K. J. (2017). Laser-assisted finishing process for improved surface finish of fused deposition modelled parts. *Journal of Manufacturing Processes*, 30, 161–177.

Thijs, L., Verhaeghe, F., Craeghs, T., Van Humbeeck, J., and Kruth, J. P. (2010). A study of the microstructural evolution during selective laser melting of Ti-6Al-4V. *Acta Materialia*, 58(9), 3303–3312.

Travitzky, N., Bonet, A., Dermeik, B., Fey, T., Filbert-Demut, I., Schlier, L., Schlordt, T., and Greil, P. (2014). Additive manufacturing of ceramic-based materials. *Advanced Engineering Materials*, 16(6), 729–754.

Wang, J., Xie, H., Weng, Z., Senthil, T., and Wu, L. (2016). A novel approach to improve mechanical properties of parts fabricated by fused deposition modeling. *Materials and Design*, 105:152–159.

Wencke, Y. L., Kutlu, Y., Seefeldt, M., Esen, C., Ostendorf, A., and Luinstra, G. A. (2021). Additive manufacturing of PA12 carbon nanotube composites with a novel laser polymer deposition process. *Journal of Applied Polymer Science*, 138, 50395.

Wilkes, J., Hagedorn, Y. C., Meiners, W., and Wissenbach, K. (2013). Additive manufacturing of ZrO_2-Al_2O_3 ceramic components by selective laser melting. *Rapid Prototyping Journal*, 19(1), 51–57.

Yadroitsev, I., Krakhmalev, P., Yadroitsava, I., Johansson, S., and Smurov, I. (2013). Energy input effect on morphology and microstructure of selective laser melting single track from metallic powder. *Journal Materials Processing Technology*, 213(4), 606–613.

11 Innovative and Hybrid Post Processes for Additively Manufactured Parts

Shahbaz Juneja[1], Jasgurpreet Singh Chohan[2], Raman Kumar[2], Arun Kumar[1], and Sushant Kumar[1]
[1]Department of Mechanical Engineering, Chandigarh University, Punjab, India
[2]Department of Mechanical Engineering, University Center for Research and Development, Chandigarh University, Punjab, India

CONTENTS

11.1 INTRODUCTION

Rapid tooling (RP) techniques that emphasize the accurate fabrication of complex geometries while reducing manufacturing time and cost make up AM technology. The traditional use of instruments, fixtures, jigs, and dies is replaced by this new generation of manufacturing techniques, also known as layer manufacturing or freeform fabrication while requiring very little maintenance [1]. Rapid prototyping and computer-aided-design (CAD) technology are combined in FDM, which makes

DOI: 10.1201/9781003276111-11

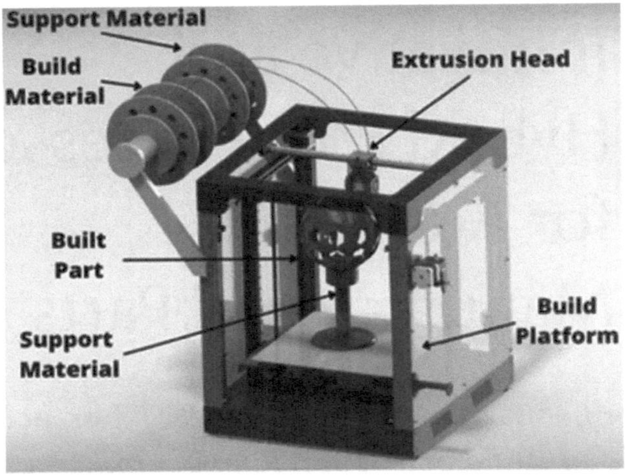

FIGURE 11.1 Schematic view of fused deposition modeling apparatus [5].

it popular because it enables the use of a variety of shapes and materials to achieve the desired properties. Many researchers are interested in FDM because it narrows the gap between product realization and conceptualization and is the most concise, adaptable, and cost-effective method with the least amount of material waste [2]. As shown in Figure 11.1, FDM uses a nozzle that moves in both *X* and *Y* directions. Extra-thin layers are accurately settled at the base without the use of fixtures, and semi-molten plastic beads are extracted from the nozzle [3, 4].

The application of rapid tooling of FDM in recent years has been aided by the demand for high quality products at lower costs and in shorter amounts of time. The latest area of rapid tooling, where less expensive castings are produced directly using the investment casting process at a lower cost, has opened up as a result of the ability to produce plastic models using FDM [6]. Rapid tooling is used to create patterns for toys, sports equipment, medical devices, automobiles, and aerospace components [7, 8]. It is an extrusion-based three-dimensional (3D) printing process that uses CAD modeling to eliminate the need for traditional tools [9]. After creating the part with CAD, an STL file is submitted, and using additive manufacturing, the desired material is loaded into a liquefier to melt and then extruded to create the final shape [10]. Due to their high portability, machinability, low by-product, and low waste levels, PLA and ABS are frequently used to create parts using the FDM method [11]. The ability to use ABS plastics in a variety of operations, such as drilling, milling, turning, and grinding, is a benefit [12].

However, there is a chance that FDM models will have some drawbacks. The roughness of a surface is one of these issues. End products need to undergo post-machining using different drilling, turning, milling, etc. to get rid of this. The staircasing effect, which makes the surface more uneven and causes the layer marks to be visible and distinguishable, is another problem that can occur during the extrusion process. The functionality of the parts is reduced by poor surface finishing, which makes them weak and ineffective [14].

Conventional and unconventional machining processing methods are used in post-processing techniques for ABS materials. Turnbull and Maxwell compared the residual stress in ABS sheets with equi-biaxial residual stress. The layer removal process led to reproducible and equi-biaxial residual stresses in the polymer sheet. In addition to this, the residual stresses determined by drilling would not be equi-biaxial and could not be balanced across specimen thickness. And finally, the most reliable results were obtained during the layer removal procedure and through thickness residual stress measurements. For assessing residual stresses in intricate geometric primitives, the hole-drilling method is very adaptable. Additionally, Meinhard et al. investigated into a hole that was conventionally drilled on a thermoplastic laminate reinforced with carbon fiber. Burrs, other than delamination, were found to be the most prevalent but non-selective type of damage in bored CFRTP after defects in laminate manufacturing and those caused by machining were identified, also grouped, and imaged. Burr formation and fiber deflection in a plasticized matrix zone, which combine with fiber cracks, are the two main damage patterns in thermoplastics [16]. Moreover, Kumar and Singh examined the delamination defect, thrust force, and surface quality of glass fiber reinforced polymer composites while drilling with various drilling tools and found that tool material and processing parameters are the primary causes of drilling defects [17]. Finally, the study demonstrates that when drilling plastic substrates using conventional processes, delamination of the layers, burr formation, and cracked edges are the major problems, leading to a transfer to non-conventional processes.

Additionally, non-conventional machining processing techniques are used for post-processing, including electro-discharge machining, abrasive water jet, laser cutting, and vapor finishing. The researcher used a laser cutting process to drill holes of varying diameters in PMMA and ABS polymer sheets. The idealized values of the optimum variable were found at 2.0 bar compressed air pressure, 500 W laser force, 0.6 m/min cutting speed, 5.0 mm workpiece laser length, 2.0 mm hole diameter, and PMMA material using cutting speed, laser power, stand-off distance, and gas pressure as independent variables. The material removal started out with the least amount of taper. When lasers are used to cut ABS polymer, the opening is more rounded at the entry than the exit, in contrast to PMMA, where the opening is more rounded at the exit than the entry [18]. Additionally, an experiment using single-pass and double-pass laser beams to machine glass fiber-reinforced material was conducted, and the results obtained demonstrate that a double beam produces a better cut finish [19]. Additionally, techniques for chemical post-processing are covered. Acetone solvent and chemical dipping are two examples of surface enhancement techniques used in these methods, which help with part machining and surface finishing [20, 21]. Drill operations carried out on 3D-printed objects. Abrasive jet machining was one of the non-conventional methods used to complete these tasks. On various components, acetone vapor drilling was done so that they could be used as fixtures, jigs, and tools, among other things. Acetone helps finish the surface, improving drill hole circularity. Acetone is a better chemical than methyl ethyl ketone and tetrahydrofuran for treating ABS parts, whether by vapor smoothing or by joining parts together. Tetrahydrofuran converts to a liquid state at high temperatures, which slows down the evaporation process despite having the same solvent properties as acetone [22, 23].

One of the main issues that must be taken into account is the significant amount of ABS substrates that were wasted during the manufacturing process and later using conventional and non-conventional processes. This waste can occur due to various reasons, including imperfect molding, trimming excess material, and discarding products that do not meet quality standards. Because ABS is not biodegradable, disposing of waste ABS is one of the biggest challenges [24]. Additionally, using improper techniques for plastic disposal endangers the ecosystem. Materials with ABS coatings contaminate soil and groundwater. Various studies on material recycling have been conducted to address this issue [25, 26].

The majority of previous work includes abrasives' drilling, laser drilling, and water jet drilling. The present study focuses on substitute methods of non-traditional drilling and machining to recycle ABS materials successfully. To minimize structural issues brought on by direct contact between tools and parts in conventional machining, waste parts are recycled in this study employing novel and hybrid post procedures, including vapor jet drilling machining and chemical processing. Vapor jet machining is a technique for drilling items that enhances the results' smoothness by using high-pressure acetone vapors.

11.2 METHODOLOGY

Recycling of ABS-used parts' manufacturing was the subject of this chapter. Acetone has been employed as a tool to improve surface finish in order to increase surface effectiveness and smoothness. The acetone drilling techniques required a result evaluation. We were able to draw conclusions about how drilling with acetone would affect 3D-printed parts as a result of the current study. The methodology for the entire process is shown below.

- Design and fabrication of hybrid chemical vapor jet machining apparatus.
- Selection and identification of process parameters of machining apparatus.
- Hybrid drilling of 3D-printed ABS parts manufactured by additive manufacturing.
- Measurement of surface roughness and circularity after machining operations.
- Summary.

11.2.1 Hybrid Experimental Set-up for Post Processing

In-house designed chemical vapor jet machine setup was used for the experiments for this study. According to Figure 11.2, the setup was split into two main sections, in the back section gauges, a compressor, and a mixing chamber are present. There was a nozzle and a reservoir in the front section, which was known as the workstation section. The flow control valve was used to control the mixture of high-pressure air and acetone that can be changed depending on the experiment's operating conditions. Compressed air pressure can be controlled using the pressure valve and evaluated using the pressure gauge mounted on the setup. The adjustable worktable can be used to change the standoff distance, which can be measured using the filler gauge.

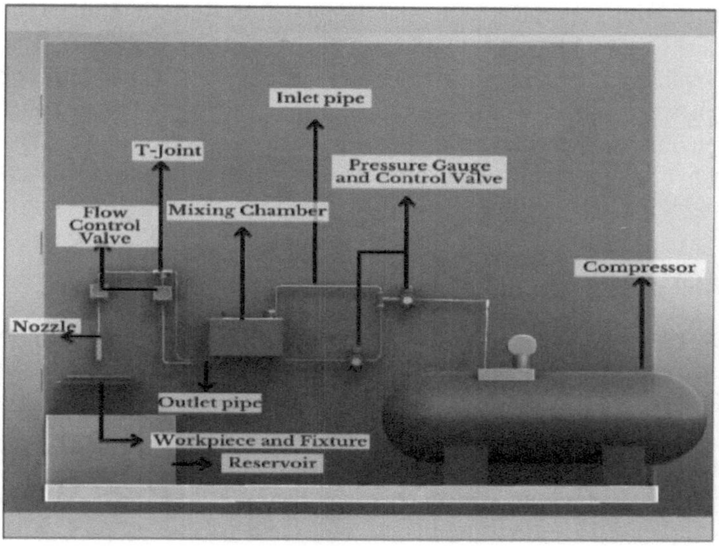

FIGURE 11.2 Experimental setup of acetone vapor jet drilling. Experimental setup (back view).

11.2.2 Work Material

ABS is a thermoplastic used to make rigid and lightweight parts. It is more practical due to its high machinability, corrosion resistance, and choice of colors, as well as its high grade strength, moisture resistance, and heat resistance. ABS materials can be used for a variety of processes and applications because of their useful temperature range of –40 °C to 100 °C. The aerospace industry, toy manufacturing, home appliance manufacturing, electrical component manufacturing, and many other industries use ABS materials.

11.2.3 Experimentation

In order to determine the impact of process variables on the circularity of the ABS workpiece and surface roughness, taking into account process constraints and the findings of the literature review, drilling with an acetone vapor jet along various variables such as standoff distance, pressure, and flow rate is chosen. These three variables were used as input parameters for acetone vapor jet machining: compressed air pressure, acetone vapor flow rate, and standoff distance. To change the distance between the nozzle and the workpiece, install a nozzle directly over the holder for the workpiece. Compressed air pressure and vapor flow rate were controlled by flow-regulating valves. Jet strikes caused the ABS jig surface to erode, leaving a hole with a smooth surface and a significant amount of circularity all the way around. The roughness of the surfaces was evaluated using the Mitutoyo Surftest SJ-410. Utilizing Vertex-311, the circularity of the drill hole was also examined to determine drilling efficiency.

The 3D-printing material ABS is not biodegradable. According to the study, there is a significant need for recycling ABS wastes and their parts. To accomplish the same result, the authors followed standard conventional practices. Two significant difficulties, though, were stress and heat generation, which resulted in internal raster damage and a rough inner surface after the molten polymer layers cooled down. This led them to employ hybrid non-traditional techniques. Considering the above study, hybrid jet drilling was employed.

11.3 FINDINGS AND DISCUSSIONS

After three tests, it was discovered that applying jet drilling to the ABS jig workpiece significantly improved both the surface roughness and the circularity. As depicted in Figure 11.2, the waste from ABS substrates has been used as a jig for the drilling procedure. Three holes were drilled in each experiment at constant pressure, flow rate, and standoff distance values of 3 bar, 15 ml/min, and 3 mm. The circularity and surface roughness of each sample were measured, and it was observed from the results that recycling of ABS parts using the acetone vapor jet drilling process is possible despite the fact that parts cannot be recycled using conventional processes because of poor surface qualities like cracks and delimitation of the processed parts. Results of surface roughness and circularity achieved after vapor jet drilling are displayed in Table 11.1

11.3.1 Analysis of Surface Roughness and Circularity

Acetone vapor is used to process the sample piece. The amount of irregularities on the top, or surface, of the sample material, is measured as its surface roughness. The abbreviation Ra, which stands for average roughness, is used to indicate surface roughness. A Mitutoyo Surftest SJ-410 is used to gauge the substrates' surface roughness. Since the material's surface has been exposed to acetone, each experiment produces a different set of effects and outcomes.

TABLE 11.1
Results of Surface Roughness and Circularity

Part No.	Hole No.	Surface Roughness	Circularity
1	1	1.391	0.3306
	2	1.368	0.3349
	3	1.407	0.3220
2	1	1.413	0.3329
	2	1.405	0.3411
	3	1.395	0.3394
3	1	1.397	0.3263
	2	1.408	0.3418
	3	1.411	0.3379

The non-treated sample has a Ra value of 9.14 m. The Ra values for the part no. 1 hole no. 1 and the part no. 2 hole no. 1 of the ABS jig treated with acetone vapor using a fixed parameter are 1.391 and 1.413 m, respectively, in comparison to the untreated sample.

Vertex 311 (Micro-InSpec Vu's) was used to measure the drilled hole circularity in order to evaluate the hole quality. According to result circularity of holes 1 and 2 in the respective parts was found to be 0.3303 and 0.3329 mm, respectively.

The material is continuously removed with the aid of the acetone-based vapor jet machining process, increasing the surface roughness value, according to the surface roughness analysis.

11.4 SUMMARY

A study has been done on the drilling effects of acetone vapor jet on the circularity and surface finishing of ABS 3D-printed parts. Following are some findings from a successful investigation:

- Surface roughness gets better over time, and recyclable materials can be treated with the acetone vapor treatment. The acetone vapor can dissolve and soften the outer layer of the plastic, creating a smoother surface finish.
- By using hybrid jet drilling, waste materials such as scrap metals and plastics can be drilled and shaped into customized jigs and fixtures. This approach offers several advantages, including cost savings, reduced waste generation, and improved sustainability.
- To focus on how chemical treatment affects the dimensional stability, hole surface quality, regularity, and smoothness of the parts post drilling.

11.5 FUTURE SCOPE

The future of additively generated parts is looking bright thanks to hybrid machining, which blends additive manufacturing and conventional machining methods. The following are some potential advantages and probable applications of hybrid machining for additive manufacturing:

1. Hybrid machining can help to improve surface smoothness and accuracy for items that are additively created by eliminating extra material and smoothing out rough surfaces. For high-precision applications like aerospace and medical devices, this can be extremely helpful.
2. Improved part functionality: By enabling the inclusion of features like threaded holes or precision bores that may not be possible with additive manufacturing alone, hybrid machining can also aid in improving the functionality of parts that have been produced using additive manufacturing.
3. Hybrid machining can also give additively generated parts more design flexibility since it enables the development of more complicated geometries that would be challenging or impossible to create using only traditional manufacturing techniques.

ACKNOWLEDGEMENT

- The authors are highly thankful to the University Center for Research and Development, Chandigarh University for the assistance.

REFERENCES

[1] A. Boschetto and L. Bottini, "Roughness prediction in coupled operations of fused deposition modeling and barrel finishing," *J. Mater. Process. Technol.*, vol. 219, pp. 181–192, 2015, doi: 10.1016/j.jmatprotec.2014.12.021.

[2] A. Peng, X. Xiao, and R. Yue, "Process parameter optimization for fused deposition modeling using response surface methodology combined with fuzzy inference system," *Int. J. Adv. Manuf. Technol.*, vol. 73, no. 1–4, pp. 87–100, 2014, doi: 10.1007/s00170-014-5796-5.

[3] A. Boschetto, V. Giordano, and F. Veniali, "Surface roughness prediction in fused deposition modelling by neural networks," *Int. J. Adv. Manuf. Technol.*, vol. 67, no. 9–12, pp. 2727–2742, 2013, doi: 10.1007/s00170-012-4687-x.

[4] P. Dudek, "FDM 3D printing technology in manufacturing composite elements," *Arch. Metall. Mater.*, vol. 58, no. 4, pp. 1415–1418, 2013, doi: 10.2478/amm-2013-0186.

[5] S. Juneja *et al.*, "Impact of process variables of acetone vapor jet drilling on surface roughness and circularity of 3D-printed ABS parts: Fabrication and studies on thermal, morphological, and chemical characterizations," *Polymers (Basel).*, vol. 14, no. 7, p. 1367, 2022, doi: 10.3390/polym14071367.

[6] D. Pal and B. Ravi, "Rapid tooling route selection and evaluation for sand and investment casting," *Virtual Phys. Prototyp.*, vol. 2, no. 4, pp. 197–207, 2007, doi: 10.1080/17452750701747088.

[7] E. Bassoli, A. Gatto, L. Iuliano, and M. G. Violante, "3D printing technique applied to rapid casting," *Rapid Prototyp. J.*, vol. 13, no. 3, pp. 148–155, 2007, doi: 10.1108/13552540710750898.

[8] S. Pattnaik, P. K. Jha, and D. B. Karunakar, "A review of rapid prototyping integrated investment casting processes," *Proc. Inst. Mech. Eng. Part L J. Mater. Des. Appl.*, vol. 228, no. 4, pp. 249–277, 2014, doi: 10.1177/1464420713479257.

[9] L. M. Galantucci, F. Lavecchia, and G. Percoco, "Quantitative analysis of a chemical treatment to reduce roughness of parts fabricated using fused deposition modeling," *CIRP Ann.–Manuf. Technol.*, vol. 59, no. 1, pp. 247–250, 2010, doi: 10.1016/j.cirp.2010.03.074.

[10] X. Peng, L. Kong, J. Y. H. Fuh, and H. Wang, "A review of post-processing technologies in additive manufacturing," *J. Manuf. Mater. Process.*, vol. 5, no. 2, 2021, doi: 10.3390/jmmp5020038.

[11] E. J. McCullough and V. K. Yadavalli, "Surface modification of fused deposition modeling ABS to enable rapid prototyping of biomedical microdevices," *J. Mater. Process. Technol.*, vol. 213, no. 6, pp. 947–954, 2013, doi: 10.1016/j.jmatprotec.2012.12.015.

[12] A. Colpani, A. Fiorentino, and E. Ceretti, "Characterization of chemical surface finishing with cold acetone vapours on ABS parts fabricated by FDM," *Prod. Eng.*, vol. 13, p. 437–447, 2019, doi: 10.1007/s11740-019-00894-3.

[13] P. M. Pandey, N. V. Reddy, and S. G. Dhande, "Improvement of surface finish by staircase machining in fused deposition modeling," *J. Mater. Process. Technol.*, vol. 132, no. 1–3, pp. 323–331, 2003, doi: 10.1016/S0924-0136(02)00953-6.

[14] D. Meinhard, A. Haeger, and V. Knoblauch, "Drilling induced defects on carbon fiber-reinforced thermoplastic polyamide and their effect on mechanical

properties," *Compos. Struct.*, vol. 256, no. October 2020, p. 113138, 2021, doi: 10.1016/j.compstruct.2020.113138.
[15] D. Kumar and K. K. Sing, "Experimental analysis of delamination, thrust force and surface roughness on drilling of glass fibre reinforced polymer composites material using different drills," in *Materials Today: Proceedings*, 2017, vol. 4, no. 8, pp. 7618–7627, doi: 10.1016/j.matpr.2017.07.095.
[16] I. A. Choudhury, W. C. Chong, and G. Vahid, "Hole qualities in laser trepanning of polymeric materials," *Opt. Lasers Eng.*, vol. 50, no. 9, pp. 1297–1305, 2012, doi: 10.1016/j.optlaseng.2012.02.017.
[17] I. A. Choudhury and P. C. Chuan, "Experimental evaluation of laser cut quality of glass fibre reinforced plastic composite," *Opt. Lasers Eng.*, vol. 51, no. 10, pp. 1125–1132, 2013, doi: 10.1016/j.optlaseng.2013.04.017.
[18] L. M. Galantucci, F. Lavecchia, and G. Percoco, "Experimental study aiming to enhance the surface finish of fused deposition modeled parts," *CIRP Ann.–Manuf. Technol.*, vol. 58, no. 1, pp. 189–192, 2009, doi: 10.1016/j.cirp.2009.03.071.
[19] C. Neff, M. Trapuzzano, and N. B. Crane, "Impact of vapor polishing on surface quality and mechanical properties of extruded ABS," *Rapid Prototyp. J.*, vol. 24, no. 2, pp. 501–508, 2018, doi: 10.1108/RPJ-03-2017-0039.
[20] A. A. Chaudhari, A. M. Godase, J. Ravindra, and N. Abhijit, "Acetone vapor smoothing: A postprocessing method for 3D printed ABS parts," *Int. J. Res. Sci. Innov.* |, vol. IV, no. V, pp. 123–127, 2017, [Online]. Available: www.rsisinternational.org.
[21] A. P. Valerga Puerta, S. R. Fernandez-Vidal, M. Batista, and F. Girot, "Fused deposition modelling interfacial and interlayer bonding in PLA post-processed parts," *Rapid Prototyp. J.*, vol. 26, no. 3, pp. 585–592, 2020, doi: 10.1108/RPJ-06-2019-0176.
[22] B. Kirby, J. M. Kenkel, A. Y. Zhang, B. Amirlak, and T. M. Suszynski, "Three-dimensional (3D) synthetic printing for the manufacture of non-biodegradable models, tools and implants used in surgery: a review of current methods," *J. Med. Eng. Technol.*, vol. 45, no. 1, pp. 14–21, 2021, doi: 10.1080/03091902.2020.1838643.
[23] S. Vyavahare, S. Teraiya, D. Panghal, and S. Kumar, "Fused deposition modelling: A review," *Rapid Prototyp. J.*, vol. 26, no. 1, pp. 176–201, 2020, doi: 10.1108/RPJ-04-2019-0106.
[24] S. Zhang *et al.*, "Non-biodegradable microplastics in soils: A brief review and challenge," *J. Hazard. Mater.*, vol. 409, p. 124525, 2021, doi: 10.1016/j.jhazmat.2020.124525.

Index

Note: Page numbers in **bold** refer to tables and those in *italic* refer to figures.